T0324779

MANAGEMENT OF HEMOSTASIS AND COAGULOPATHIES FOR SURGICAL AND CRITICALLY ILL PATIENTS

MANAGEMENT OF HEMOSTASIS AND COAGULOPATHIES FOR SURGICAL AND CRITICALLY ILL PATIENTS

An Evidence-Based Approach

ANDY NGUYEN, MD
Professor of Pathology and Laboratory Medicine
University of Texas-Houston Medical School

AMITAVA DASGUPTA, PhD
Professor of Pathology and Laboratory Medicine
University of Texas-Houston Medical School

AMER WAHED, MD
Associate Professor of Pathology and Laboratory Medicine
University of Texas-Houston Medical School

ELSEVIER

AMSTERDAM • BOSTON • HEIDELBERG • LONDON • NEW YORK • OXFORD
PARIS • SAN DIEGO • SAN FRANCISCO • SINGAPORE • SYDNEY • TOKYO

Elsevier
Radarweg 29, PO Box 211, 1000 AE Amsterdam, Netherlands
The Boulevard, Langford Lane, Kidlington, Oxford OX5 1GB, UK
50 Hampshire Street, 5th Floor, Cambridge, MA 02139, USA

Notices

Knowledge and best practice in this field are constantly changing. As new research and experience broaden
our understanding, changes in research methods, professional practices, or medical treatment may become necessary.

Practitioners and researchers must always rely on their own experience and knowledge in evaluating and using
any information, methods, compounds, or experiments described herein. In using such information or methods they
should be mindful of their own safety and the safety of others, including parties for whom they have a
professional responsibility.

To the fullest extent of the law, neither the Publisher nor the authors, contributors, or editors, assume any liability for
any injury and/or damage to persons or property as a matter of products liability, negligence or otherwise, or from any
use or operation of any methods, products, instructions, or ideas contained in the material herein.

Medicine is an ever-changing field. Standard safety precautions must be followed, but as new research
and clinical experience broaden our knowledge, changes in treatment and drug therapy may become
necessary or appropriate. Readers are advised to check the most current product information provided by the
manufacturer of each drug to be administered to verify the recommended dose, the method and duration of
administrations, and contraindications. It is the responsibility of the treating physician, relying on experience
and knowledge of the patient, to determine dosages and the best treatment for each individual patient.
Neither the publisher nor the authors assume any liability for any injury and/or damage to persons or property
arising from this publication.

Library of Congress Cataloging-in-Publication Data
A catalog record for this book is available from the Library of Congress

British Library Cataloguing-in-Publication Data
A catalogue record for this book is available from the British Library

ISBN: 978-0-12-803531-3

For Information on all Elsevier books,
visit our website at http://www.elsevier.com/

  Working together
to grow libraries in
developing countries

www.elsevier.com • www.bookaid.org

Publisher: Mara Connor
Acquisition Editor: Tari Broderick
Editorial Project Manager: Jeffrey Rossetti
Production Project Manager: Melissa Read
Designer: Maria Ines Cruz

Typeset by MPS Limited, Chennai, India

CONTENTS

PREFACE

Transfusion medicine is an important aspect of clinical medicine because it is a lifesaving procedure. The clinical laboratory plays an important role in transfusion medicine because it is in this laboratory that crossmatching and other tests necessary for successfully transfusing a patient are conducted. There are many excellent textbooks and reference books on transfusion medicine. This book focuses on a subspeciality of transfusion medicine in which management of hemostasis and coagulopathies in a surgical or critically ill patient is approached from the standpoint of evidence-based medicine, where transfusion is guided by coagulation status of a patient based mostly on laboratory test results. In our institution, pathologists are involved in managing perioperative bleeding of patients after cardiac surgery, determining the transfusion needs of patients based on laboratory test results (hemotherapy service). A special algorithm was developed by Dr Andy Nguyen and his group at the University of Texas at Houston, and this book describes our experience in successfully applying this algorithm for better patient management as well as reducing wastage of expensive blood products. This approach is unique, and we believe this book will help clinicians and clinical pathologists to adopt such practice in their hospitals.

This book will appeal to a broad section of medical and nurse practitioners, from internal medicine to surgery and anesthesiology. For general medicine and surgery practitioners, this book will provide practical guidelines on how to control bleeding and use blood products rationally for best medical practice. For pathologists who are involved in laboratory medicine and transfusion medicine, this book will help them understand proper utilization of blood product and how to avoid unnecessary waste, which eventually will save millions of dollar for hospitals.

We thank Robert Hunter, MD, PhD, and chair of our department, for encouraging us to write this book, which presents a unique approach for pathologists to practice clinical medicine not just as consultants but also as providers of the service. We also thank our wives for supporting us during the long hours we spent preparing the manuscript. Our readers will be the final judge of the utility of this book. If readers think that the information provided in this book is clinically useful and can adopt this knowledge in their clinical practice, our efforts will be duly rewarded.

Andy Nguyen
Amitava Dasgupta
Amer Wahed

CHAPTER 1

Coagulation-Based Tests and Their Interpretation

Contents

1.1 INTRODUCTION

Blood clotting (coagulation) is initiated within seconds after vascular injury to counteract bleeding. The main purpose of coagulation is to seal an injured vessel, which is accomplished by aggregation of platelets at the site of injury. Antimicrobial peptides are released from platelets when coagulation is activated. In addition, an intact platelet—fibrinogen plug can provide an active surface that allows the recruitment, attachment and

A. Nguyen, A. Dasgupta, A. Wahed:
Management of Hemostasis and Coagulopathies for Surgical and Critically Ill Patients.
DOI: http://dx.doi.org/10.1016/B978-0-12-803531-3.00001-X

activation of phagocytosing cells. Various tests are performed prior to surgery to assess risk of bleeding and during the perioperative and post-operative period to guide transfusion using appropriate blood components.

Commonly used coagulation tests for such purposes include the following:

- Complete blood count
- Tests for platelet function
- Disseminated intravascular coagulation (DIC) screen, which consists of prothrombin time (PT), partial thromboplastin time (PTT), thrombin time (TT), fibrinogen level and D-dimer
- Antithrombin III level
- Thromboelastograph (TEG)
- Special testing such as anti-Xa assay, diluted thrombin time and tests for heparin-induced thrombocytopenia

1.2 COMPLETE BLOOD COUNT

A complete blood count (CBC) is one of the most common laboratory tests ordered by clinicians, and this test is also essential prior to surgery to evaluate a patient for bleeding risk. For CBC analysis, the specimen must be collected in an ethylenediaminetetraacetic acid (EDTA) tube (lavender or purple top).

CBC consists of white blood cell (WBC) count (or leukocyte count), WBC differential count, red blood cell count (or erythrocyte count), hematocrit, hemoglobin, mean corpuscular volume, mean corpuscular hemoglobin, mean corpuscular hemoglobin concentration, red cell distribution width, platelet count and mean platelet volume. Some of these parameters are directly measured by the analyzer, whereas others are calculated from various measured parameters. In general, hemoglobin and hematocrit levels are used to determine the need for and extent of packed red blood cell (PRBC) transfusion. Platelet count is used as an important parameter to determine the necessity for platelet transfusion.

1.3 TESTS FOR PLATELET FUNCTION

Available analyzers for various tests for platelet function include PFA-100, Plateletworks and VerifyNow. In addition, the platelet aggregation test is also useful. These tests are used to determine platelet function and to decide if platelet transfusion may be needed.

Table 1.1 Interpretation of PFA-100

Result	CEPI closure time	CADP closure time
No platelet dysfunction	Normal	Normal
vWD 2N, clopidogrel, ticlopidine, storage pool disease	Normal	Normal
Aspirin effect	Prolonged	Normal
Platelet dysfunction or vWD	Normal or prolonged	Prolonged

vWD 2N: von Willebrand disease type 2N.

1.3.1 PFA-100

The PFA-100 system is a platelet function analyzer that may be used to assess platelet-related primary hemostasis. The instrument utilizes two disposable cartridges that are coated with platelet agonists. Blood is collected in a citrate tube and is then transferred to a sample cup. Blood is aspirated from the sample cup and passes through an aperture in a membrane that is coated with platelet agonists. When platelet aggregation takes place, the aperture closes and the blood flow stops. The time taken for the aperture to close is the closure time. One membrane is coated with collagen/epinephrine (CEPI), and the other membrane is coated with collagen/adenosine diphosphate (CADP). If the CEPI closure time is prolonged but the CADP closure time is normal, this is most likely due to aspirin. If both CEPI and CADP closure times are prolonged or CEPI closure time is normal and CADP closure time is abnormal, this implies platelet dysfunction or von Willebrand disease. PFA-100 is insensitive to von Willebrand disease type 2N, clopidogrel, ticlopidine and storage pool disease. PFA-100 results are affected by thrombocytopenia and low hematocrits. They are not affected by heparin or deficiencies of clotting factors other than fibrinogen. Interpretation of PFA-100 results is summarized in Table 1.1.

1.3.2 VerifyNow

VerifyNow is a rapid, turbidimetric whole blood assay capable of evaluating platelet aggregation. This assay is based on the ability of activated platelets to bind with fibrinogen. The VerifyNow IIb/IIIa assay utilizes fibrinogen–coated microparticles, whereas thrombin receptor-activating peptide is used as an agonist to maximally stimulate platelets in order to determine platelet function. If GpIIb/IIIa (glycoprotein IIB/IIIa) antagonists are present in patients' blood, then platelet aggregation should be reduced. In the VerifyNow aspirin assay, arachidonic acid (AA) is used

Table 1.2 Interpretation of VFN results

Result	Conclusion
ARU >550	Absence of clinically significant effect of aspirin
ARU <550	Presence of clinically significant effect of aspirin
PRU >210	Absence of clinically significant effect of clopidogrel
PRU <210	Presence of clinically significant effect of clopidogrel

as the agonist in order to measure the antiplatelet effect of aspirin. Results are expressed in aspirin reactive units (ARU). The VerifyNow P2Y12 assay is similarly used to assess the antiplatelet effect of clopidogrel, and results are expressed in Plavix reactive units (PRU). VerifyNow assays may be used to assess efficacy of the previously mentioned drugs, check patient compliance and assess residual effects of these drugs if a patient is scheduled to undergo surgery or invasive procedures. Interpretation of VFN results is outlined in Table 1.2.

1.3.3 Plateletworks

Plateletworks is a test for the assessment of platelet function using whole blood. For this test, kits as well as an impedance cell counter (ICHOR II analyzer) are commercially available from the Helena Laboratory so that this test can be used as a point-of-care test. This test assesses platelet function by comparing the platelet count before and after exposure with a specific platelet agonist. For this test, blood is collected in EDTA tubes as well as in other platelet agonist-containing tubes, such as ADP, AA and collagen. In the agonist tube, functional platelets should aggregate and the nonfunctional platelets should not aggregate. Then a hematology analyzer such as ICHOR II is used to count the number of platelets in the EDTA tube and also the number of unaggregated platelets in the agonist tube. The unaggregated platelets are dysfunctional. In order to calculate the number of functional platelets, the platelet count in the presence of a platelet agonist should be subtracted from the platelet count obtained from using blood collected in the EDTA tube. Plateletworks is useful in monitoring platelet response in patients receiving antiplatelet agents, including aspirin and clopidogrel.

1.3.4 Platelet Aggregation

Platelet aggregation test using platelet aggregometry is a widely used laboratory test to screen patients with inherited or acquired defects of platelet function. Platelet aggregometry measures the increase in light transmission through platelet-rich plasma that occurs when platelets are aggregated due to the addition of an agonist. For this test, blood should be collected in a citrate tube, and the test should be performed within 4 h of blood collection. Prior to analysis, specimen should be stored at room temperature. Platelet-rich plasma is obtained from the sample by centrifugation. Ideally, the platelet count of the platelet-rich plasma should be approximately 200,000–250,000; if the platelet count is higher, it can be adjusted by saline.

Prior to actual testing, the platelet-rich plasma should be left at room temperature for approximately 30 min because the test is performed at 37°C. If the original platelet count of the patient is less than 100,000, then the test might be invalid. If the test needs to be performed, then the platelet count of the control should also be lowered. Various agonists used for this test include AA, collagen, ristocetin, ADP and epinephrine. Platelet aggregation can also be performed using whole blood instead of platelet-rich plasma. In general, agonists such as ADP and epinephrine are considered weak. These two weak agonists, in low concentration in a normal person, demonstrate two waves of aggregation. The primary wave of aggregation is due to activation of the GpIIb/IIIa receptor. The secondary wave is due to platelet granule release. Lack of secondary wave implies a storage pool disorder due to a reduced number of granules or defective release of granule contents. The use of these agonists (ADP and epinephrine) in higher concentrations results in the two waves merging into one wave of aggregation. Collagen characteristically demonstrates an initial shape change before the wave of aggregation. This is seen as a transient increase in turbidity. Effective aggregation is typically considered as 70–80% aggregation. However, values should be compared with the control. Values of 60% or greater are generally considered to be adequate. Note that samples for platelet aggregation should be walked into a clinical laboratory and should not be sent by a pneumatic tube system [1]. Patterns of platelet aggregation include the following:

- Normal: Adequate aggregation with ADP, collagen, epinephrine, AA and ristocetin at high dose but not at low dose (6 mg/mL or less).
- von Willebrand/Bernard–Soulier pattern: Adequate aggregation with ADP, collagen, epinephrine and AA but not at higher dose of ristocetin.
- von Willebrand type IIB pattern/pseudo-von Willebrand pattern: Increased aggregation with low dose of ristocetin.

- Glanzmann's thrombasthenia/hypofibrinogenemia pattern: Adequate aggregation with ristocetin but impaired aggregation with all other agonists is observed. Uremia and multiple antiplatelet medications (eg, aspirin and Plavix together) can also produce similar results.
- Disorder of activation (storage pool disorder): Loss of secondary wave of aggregation with ADP and epinephrine at lower doses is the characteristic of this disorder. Storage pool disease is the most common inherited platelet function defect.
- Aspirin effect: Significant impairment with aggregation with AA is observed in patients taking aspirin. There may be impairment of aggregation (not as much as with AA) with ADP, collagen and epinephrine.
- Plavix (clopidogrel) effect/ADP receptor defect: Significant impairment of aggregation with ADP. We use an ADP concentration of 50 μM/mL to assess the effect of clopidogrel on platelets. If the extent of aggregation is 70—80% with this concentration of ADP, it may be concluded that there is no significant clopidogrel effect. Interpretation of platelet aggregation is summarized in Table 1.3.

1.4 PROTHROMBIN TIME

PT is a widely used test to evaluate secondary hemostasis. In this test, platelet-poor plasma from a patient (collected in a blood collection tube containing sodium citrate) is mixed with thromboplastin and calcium and then clotting time is determined at 37°C using a variety of methods, including photooptical and electromechanical detection. Automated coagulation analyzers are commercially available for measuring PT along with other coagulation parameters. PT is a functional measure of the extrinsic pathway and common pathway, and the reference range is 8.8—11.6 s. Therefore, PT is a useful test to detect inherited or acquired defects in coagulation related to extrinsic pathway. However, often PT is reported as international normalized ratio (INR). The thromboplastin used may vary from laboratory to laboratory and from country to country. However, reporting results as INR ensures results are comparable between different laboratories. The INR is calculated as follows:

$$INR = [Patient\ PT/Mean\ normal\ PT]^{ISI}$$

Table 1.3 Interpretation of platelet aggregation results

Result	Aggregation with high-dose Ristocetin	Aggregation with low-dose Ristocetin	Aggregation with high-dose ADP	Aggregation with low-dose ADP	Aggregation with collagen	Aggregation with arachidonic acid	Aggregation with epinephrine
Normal	Adequate	No aggregation	Adequate	Adequate	Adequate	Adequate	Adequate
vWD/Bernard–Soulier disease	Impaired aggregation	No aggregation	Adequate	Adequate	Adequate	Adequate	Adequate
vWD type IIB/pseudo-platelet pattern	Adequate	Aggregation present	Adequate	Adequate	Adequate	Adequate	Adequate
Glanzmann's pattern (also seen in hypofibrinogenemia, uremia and multiple antiplatelet medications)	Adequate	No aggregation	Impaired aggregation	Impaired aggregation	Impaired aggregation	Impaired aggregation	Impaired aggregation
Aspirin effect	Adequate	No aggregation	Adequate	Mild to moderately impaired aggregation	Mild to moderately impaired aggregation	Markedly impaired aggregation	Impaired aggregation
Plavix effect	Adequate	No aggregation	Impaired aggregation	Impaired aggregation	Mild to moderately impaired aggregation	Mild to moderately impaired aggregation	Impaired aggregation
Storage pool disorder	Adequate	No aggregation	Loss of secondary wave	Loss of secondary wave	Adequate	Adequate	Loss of secondary wave

where ISI is the international sensitivity index, which is available from the reagent package insert. The normal INR range is 0.8−1.2. Causes of prolonged PT include the following:

- Coumarin (warfarin)
- Vitamin K deficiency (dietary deficiency or failure of absorption)
- Factor deficiency (inherited or acquired; eg, liver disease) or factor inhibitor of the extrinsic and common pathway

1.5 PARTIAL THROMBOPLASTIN TIME

PTT (also known as activated partial prothrombin time (aPTT)) is another useful test for evaluation of secondary homeostasis. In this test, the patient's platelet-poor plasma (citrated plasma but oxalate can also be used), surface activating agent (silica, kaolin, celite or ellagic acid), calcium and platelet substitute (crude phospholipid) are mixed and clotting time is determined most likely using an automated coagulation analyzer. It is a functional measure of the intrinsic pathway as well as common pathway and can detect hereditary or acquired defects of the coagulation factors XII, XI, X, IX, VIII, V, prothrombin and fibrinogen. PTT or aPTT is called partial thromboplastin time because of the absence of tissue factor (thromboplastin) in the tests, and the normal value varies from laboratory to laboratory but is usually between 25 and 39 s. Causes of prolonged PTT include the following:

- Heparin, direct thrombin inhibitors (DTIs)
- Factor deficiency (inherited or acquired) of the intrinsic and common pathway
- Inhibitors: VIII and IX inhibitors, lupus anticoagulant or lupus antibody (LA)
- von Willebrand disease
- HMWK (Fitzgerald factor) deficiency
- Pre-kallikrein (Fletcher factor) deficiency
- Spurious causes

Spurious causes of prolonged PTT include high hematocrit; underfilling of the citrate tube; and EDTA contamination, which may occur if the purple-top tube is collected before the blue-top tube. Delay in transport and processing (testing should be done within 4 h of collection; sample should be stored at room temperature) may also cause spurious results. Interpretation of PT and PTT results is summarized in Table 1.4.

Table 1.4 Interpretation of PT and PTT results in various clinical scenarios

PT result	PTT result	Clinical scenario
Normal	Prolonged	Factor deficiency of the intrinsic pathway, above the common pathway (eg, factor VIII, factor IX deficiency) Heparin therapy, DTI therapy von Willebrand disease Inhibitors (eg, lupus anticoagulant, factor VIII or IX inhibitor)
Prolonged	Normal	Liver disease, vitamin K deficiency, warfarin therapy
Prolonged	Prolonged	Factor deficiency of the common pathway Heparin therapy (high dose), DTI therapy Warfarin therapy (high dose) Lupus anticoagulant (strong antibody) DIC

1.6 THROMBIN TIME

TT is a test in which the patient's plasma is mixed with thrombin and clotting time is determined. It is a measure of functional fibrinogen. Heparin produces prolonged TT. Patients on heparin, however, have a normal reptilase time. Causes of prolonged TT include the following:

- Heparin
- Hypofibrinogenemia
- Dysfibrinogenemia
- Thrombolytic therapy

1.7 FIBRINOGEN LEVEL

The most commonly performed fibrinogen assay is a modified thrombin test known as the Clauss fibrinogen assay. It is used to measure the qualitative amount of fibrinogen. Heparin and DTIs typically do not interfere with the assay because testing is performed with diluted sample, effectively diluting out thrombin inhibitors.

1.7.1 D-Dimer

D-Dimer is a fibrin degradation product. It is named as such because it contains two cross-linked D fragments of the fibrin protein. Increased levels imply increased fibrinolysis and can be seen in DIC and thrombotic states.

1.7.2 Antithrombin III Level

Antithrombin inactivates factor Xa (activated factor X) and activated factor II (thrombin). Therefore, activity is increased by heparin. In surgeries in which cardiopulmonary bypass is used and heparin is the anticoagulant employed, measuring antithrombin III (ATIII) levels prior to surgery may be useful. Individuals with low ATIII levels may require additional heparin to achieve desired anticoagulated state. This is termed "heparin resistance." Heparin may also be stored in adipose tissue and be released back into the circulation several hours after surgery. This is referred to as rebound heparin. This may result in bleeding. Both heparin resistance and rebound heparin can be avoided if low ATIII levels are corrected prior to surgery.

Low levels of ATIII levels may be seen in the following:
- Inherited deficiency
- Liver disease
- Nephrotic syndrome
- DIC
- L-Asparaginase therapy reduces hepatic synthesis of ATIII.

1.7.3 Mixing Study

Individuals with prolonged PT or prolonged PTT or both may undergo PT/PTT mixing study. The objective of this test is to determine if prolonged PT or PTT is due to a factor deficiency or due to the presence of inhibitor. In this test, the patient's plasma is mixed with an equal volume of normal plasma, and PT or PTT or both are measured at 0 and 1−2 h. Failure of correction of prolonged PT/PTT indicates the presence of inhibitors. If results of PTT at 0 and 1−2 h are similarly prolonged, it implies lupus anticoagulant. If results show time-dependent prolongation, it implies coagulation factor antibody (typically factor VIII inhibitor). If PT and PTT are both prolonged and mixing study shows correction, then most likely there is deficiency of factor in the common pathway. Hemophiliacs typically have prolonged PTT with normal PT. Mixing study should show correction. Note that patients on heparin or DTIs will have prolonged PTT and TT. In such situations, the mixing study is invalid and needs to be repeated once the patient is off these anticoagulants.

1.8 THROMBOELASTOGRAPHY

The whole blood thromboelastography (TEG) is a method of assessing global hemostasis and fibrinolytic function that includes interaction of primary and secondary hemostasis, and subsequently, defect in one component of hemostasis can affect the other to a certain extent. This technique has existed for more than 60 years, but improvements in technology have led to increased utilization of this test in clinical practice for monitoring hemostatic and fibrinolytic rearrangements [2]. TEG is a visualization of viscoelastic changes that occur during in vitro coagulation and provides a graphical representation of the fibrin polymerization process; however, during interpretation of TEG data/tracing, it is important to focus on the most significant defect. In classical TEG, a small sample of blood (typically 0.36 mL) is placed into a cuvette (cup), which is rotated gently through 4°45′ (cycle time 6/min) to imitate sluggish venous flow and activate coagulation. When a sensor shaft is inserted into the sample, a clot forms between the cup and the sensor. The speed and strength of clot formation are measured in various ways but typically computing the speed at which a specimen coagulates depends on various factors, including the activity of the plasmatic coagulation system, platelet function, fibrinolysis and other factors that can be affected by illness, environment and medications.

TEG analysis provides five basic parameters: R (reaction time), K value, angle alpha and maximum amplitude (MA) and Ly30 (clot lysis at 30 min). Reaction time is measured in seconds and represents initial latency from start of the test until the initial fibrin formation (usually amplitude of 2 mm). K value is also measured in seconds and indicates time taken to achieve a certain level of clot strength (usually amplitude of 20 mm). Alpha angle (degrees) measures the speed of fibrin buildup and cross-linking taking place, thus assesses the rate of clot formation. MA (measured in millimeters) represents the ultimate strength of the fibrin clot. Possibly the most important information provided by the TEG is clot strength, which may help to resolve whether the bleeding is related to coagulopathy or a mechanical bleeding. Clot strength is measured by MA value; a low MA value indicates platelet dysfunction. G is a computer-generated value reflecting the strength of the clot from initial fibrin blast to fibrinolysis:

$$G = (5000 \times Amplitude)/100 - Amplitude \ (Normal \ 5.2-12.4)$$

Therefore, MA and G value represent a direct function of the maximum dynamic properties of fibrin and platelet bonding via GpIIb/IIIa and represent the ultimate strength of the fibrin clot. However, G is the best measurement of clot strength [3]. Clot index (CI) represents

Table 1.5 Various parameters of TEG

Parameter	Comment
R: Reaction (measured in seconds)	The value indicates the time until the first evidence of a clot is detected
K: Clot kinetics (measured in seconds)	K value is the time from the end of R value until the clot reaches 20 min, and this value represents the speed of clot formation
Angle alpha (measured in degrees)	Measures the rapidity of fibrin buildup and cross-linking (clot strengthening) and is the tangent of the curve made as the K is reached
MA: Maximum amplitude (measured in millimeters)	Direct measure of highest point of the TEG curve and represents clot strength
G: Measures clot strength	G is calculated from MA
CI: Clotting index	CI is a mathematic equation calculated from R, K, alpha angle and MA values
Ly30: Clot lysis at 30 min	This provides information on the fibrinolytic activity during the first 30 min after MA and is a calculated value

hemostasis profile and is calculated based on R, K, alpha angle and MA. Ly30 indicates percentage decrease in amplitude at 30 min after MA the stability or degree of fibrinolysis. Various parameters obtained from TEG analysis are listed in Table 1.5.

TEG may be used to dictate use of blood products and certain medications in a bleeding patient. Blood products that may be used in bleeding patients include PRBCs, fresh frozen plasma, cryoprecipitate and platelets. Medications that may be used include protamine, antifibrinolytic agents (eg, epsilon amino caproic acid and tranexamic acid). However, use of PRBCs is dictated by hemoglobin and hematocrit, and analysis TEG is not used for this purpose. Use of FFP is indicated if the R time is prolonged. In such patients, PT and PTT should also be prolonged.

Cryoprecipitate is used in bleeding patients with hypofibrinogenemia and uremic thrombocytopathia. Cryoprecipitate is also useful in von Willebrand disease patients who are bleeding. In hypofibrinogenemia, the angle alpha is low with a normal MA. These patients also have prolonged TT and low levels of fibrinogen. With regard to the use of platelets, platelet transfusion is valuable in thrombocytopenia and thrombocytopathia, and in both conditions MA should be low. Complete blood count will also document thrombocytopenia.

Protamine is used to neutralize heparin, and TEG must be performed with and without heparinase (an enzyme that destroys heparin). If the R value in the TEG analysis without heparinase is greater than 50% longer than the R with heparinase, protamine is indicated. In these patients, TT and PTT are also prolonged, but PT is not as prolonged as PTT with heparin.

Fibrinolysis may be primary or secondary. Causes of primary fibrinolysis include physiological (due to thrombolytic therapy) or pathological causes. Secondary fibrinolysis is seen in DIC. In primary and secondary hyperfibrinolysis, Ly30 is prolonged. However, in primary fibrinolysis, the MA value from the TEG analysis is low, as is the CI value. In secondary fibrinolysis, MA is normal or high with high CI. Antifibrinolytic agents are only used in patients with primary pathologic fibrinolysis. It is very important to note that TEG values are normal in patients on aspirin and clopidogrel.

1.8.1 Platelet Mapping

Platelet mapping is a special TEG assay to measure the effects of antiplatelet drug therapy on platelet function. Antiplatelet drugs, whose efficacy can be tested, include the following:
- ADP receptor inhibitors such as clopidogrel and ticlopidine
- Arachidonic acid pathway inhibitors such as aspirin
- GpIIb/IIIa inhibitors such as abciximab, tirofiban and eptifibatide

The Platelet Mapping assay specifically determines the MA reduction present with antiplatelet therapy and reports the percentage inhibition and aggregation.

The Platelet Mapping assay measures platelet function in the presence of antiplatelet drugs in a patient's blood sample. The results obtained by the TEG 5000 analyzer will be reported as percentage inhibition and percentage aggregation. The Platelet Mapping assay measures the presence of platelet-inhibiting drugs using whole blood and the following four different steps:
- No additive sample to measure total platelet function and the contribution of fibrin to MA and yields $MA_{Thrombin}$. Thrombin overrides inhibition at other platelet activation pathways and indicates complete activation.
- Activator F is added to measure the contribution of fibrin only to MA and yields MA_{Fibrin}, which indicates no activation.

- Activator F is added to the sample along with ADP to measure MA due to ADP receptor uninhibited platelets to yield the value of MA_{ADP}, which indicates activation of noninhibited platelets.
- Activator F is added to the sample along with AA to measure MA due to TxA2 pathway to yield the value of MA_{AA}, which also indicates activation of noninhibited platelets.

The presence of platelet-inhibiting drugs is reflected in a reduction in MA values. The percentage inhibition is derived by the following equation:

$$\text{Percent MA Reduction} = 100 - [\{(MA_P - MA_F)/(MA_T - MA_F)\} \times 100]$$

where MA_P represents MA_{ADP} or MA_{AA}, MA_F represents MA_{Fibrin} and MA_T represents $MA_{Thrombin}$. The optimum time, from the time the blood enters the syringe to the time it is placed in the TEG instrument, is 4 min.

1.8.2 Anti-Xa Assay

The plasma anti-Xa assay may be used to monitor patients on unfractionated heparin (UFH) or low-molecular-weight heparin (LMWH). The activity of both UFH and LMWH is dependent on binding to antithrombin (AT), which results in increased inhibitory activity of AT. LMWH has primarily anti-Xa activity, and UFH has both anti-Xa and anti-IIa activity. The anti-Xa activity of AT can be measured by a clotting-based or chromogenic assay.

1.8.3 Plasma-Diluted Thrombin Time

This is a test consisting of 1 part patient plasma diluted with 3 parts normal plasma. This test has been shown to blunt the sensitivity of TT and be more accurate for drug monitoring. Patients who have elevated PTT values at baseline (eg, LA) and are put on DTIs may be monitored by this test.

1.9 HEPARIN-INDUCED THROMBOCYTOPENIA

There are two types of heparin-induced thrombocytopenia (HIT):
- HIT I: Here, a decline in platelet count within the first 2 days of heparin administration is observed. This disorder is due to the direct effect of heparin on platelet activation. There is no need to discontinue heparin therapy because this effect is nonimmune in nature.

- HIT II: This affects 0.2—5% of patients who receive heparin for more than 4 days. Clinical features include heparin administration for more than 4 days with reduction of platelet count by 50% or greater. Patients who are at risk for HIT II are females, surgical patients (especially cardiac and orthopedic surgery), patients with previous exposure to heparin and patients receiving UFH. In HIT II, antibodies are formed that bind to platelet factor-4 (PF-4) and result in platelet aggregation causing thrombosis. It is thought that the antibodies are initially formed against antigens mimicking the PF-4/heparin complex (eg, PF-4 complex with polysaccharide on the surface of bacteria). Tests for HIT include the following:

- Enzyme-linked immunosorbent assay (ELISA)-based test: Heparin PF-4 complexes are coated on a plate to detect antibody in the patient's plasma or serum. The test is very sensitive but with low specificity. Results are published as positive or negative based on the optical density (OD) value of the patient compared to the control. It is important to appreciate that individuals with positive ELISA do not necessarily have platelet-activating antibody. However, individuals with OD >1.0 are more likely to be associated with platelet-activating antibody.

- Heparin-induced platelet aggregation: Heparin-induced platelet aggregation is performed using donor platelets and the patient's serum. The control test has no heparin added to it. Another test will be performed with a low concentration of heparin being added to it (0.1—0.3 U/mL). One more test will be performed with a high concentration of heparin added to it (10—100 U/mL). A positive test is one in which there is platelet aggregation (>25%) with a low concentration of heparin, but not with a high concentration or in the control.

- Serotonin release assay: Platelets from a normal donor are incubated with radiolabeled serotonin (^{14}C), and then labeled platelets are incubated with the patient's serum in the absence and presence of heparin in therapeutic (0.1 U/mL) or high concentration of heparin (100 U/mL). For a test to be positive, there must be at least a 20% radioactivity release in the presence of 0.1 U/mL of heparin, but release of radioactivity must be substantially reduced in the presence of high heparin concentration. Serotonin release assay is considered the gold standard. However, this is often a send-out test and thus results are available after considerable delay.

1.9.1 Bleeding Patient With Normal Coagulation Tests

A patient who is bleeding may show normal values of PT/PTT. This situation may be due to the following conditions:

- Thrombocytopenia and thrombocytopathia
- Factor XIII deficiency (clot is not soluble in 5 M urea solution for 24 h)
- Increased primary fibrinolysis (eg, α_2-antiplasmin deficiency)

1.10 CONCLUSIONS

Coagulation testings are essential for evaluating bleeding patients, including patients scheduled for surgery or during postsurgical care. The VFN test is useful in evaluating bleeding risk of a patient scheduled for surgery, especially if the patient is on aspirin or Plavix. TEG is a method that has been used for more than 60 years, but with modern modification of the technique and the availability of automated analyzers, TEG analysis is useful in evaluating the bleeding state of a patient, although values may remain normal in some patients with moderate coagulopathy. Correlation of TEG values with DIC screen is optimal.

REFERENCES

[1] Tiffany ML. Technical considerations for platelet aggregation and related problems. Crit Rev Clin Lab Sci 1983;19:27–69.
[2] Chen A, Teruya J. Global hemostasis testing thromboelastography: old technology, new application. Clin Lab Med 2009;29:391–407.
[3] da Luz LT, Nascrimento B, Rizoli S. Thrombelastography (TEG): practical considerations and its clinical use in trauma patients. Scand J Trauma Resusc Emerg Med 2013;21:29.

CHAPTER 2

Blood Bank Testing and Blood Products

Contents

2.1 INTRODUCTION

Several tests are routinely performed in blood banks, including the following:

- ABO/Rh typing
- Antibody screen
- Crossmatch

A. Nguyen, A. Dasgupta, A. Wahed:
Management of Hemostasis and Coagulopathies for Surgical and Critically Ill Patients.
DOI: http://dx.doi.org/10.1016/B978-0-12-803531-3.00002-1

ABO typing (blood group typing; A, B, AB and O) is usually performed using two methods, and the conclusion of both methods should be the same. If not, additional testing is required to resolve the discrepancy. Antibody screen is done to check for the presence of preformed antibodies against red blood cell (RBC) antigens. These antibodies of the patient may react with antigens of the transfused RBCs and cause hemolysis. If the antibody screen is positive, then additional tests will be performed to identify the antibody.

If the blood bank is a donor center, then there are several tests also performed on the donor's blood prior to release of blood unit for general use. These tests include screening for transmissible infectious diseases. Donor blood is tested for hepatitis B, hepatitis C, HIV-1 and HIV-2, as well as human T-cell lymphotropic virus (HTLV-I) and HTLV-II, *Treponema pallidum*, *Trypanosoma cruzi* and West Nile virus. If the donor blood shows any positive test, then the blood unit is discarded. If all test results are negative, the blood may be released for transfusion.

During transfusion, if there is any evidence or suspicion of transfusion reaction, then the transfusion is discontinued immediately and the clinical team initiates a transfusion reaction workup. The blood components that are used for transfusion are returned to the blood bank. Then, the blood bank professionals investigate the issue by performing certain tests.

In order to provide safe blood and avoid preventable risks associated with blood transfusions, all routine transfusion medicine services, starting with blood donation, testing, processing, providing blood and blood products and ending with administration and reporting adverse events, are highly regulated by the Code of Federal Regulations and American Association for Blood Banks (AABB).

2.2 BLOOD COLLECTION AND TESTINGS

Misidentification of patients during the collection of blood for blood bank testing may have serious consequence, including death. Therefore, special care must be taken and a strict protocol must be followed during blood collection. Blood sample collection should be performed directly by the phlebotomist/nurse after correctly identifying the patient by full name and hospital identification number (attached wristband). The blood sample should be labeled at bedside. Tubes should not be prelabeled. Each patient has a red arm band with a unique blood bank identification number. This number should be included as part of the labeling of the tube.

Acceptable tubes for blood bank testing are siliconized plain tubes (red top) with no additives or tubes with potassium ethylenediaminetetraacetic acid (K_2EDTA) as anticoagulant (purple top). Because purple-top tubes are also used for complete blood count (CBC), pink-top tubes (K_3EDTA) with the same anticoagulant may be used for samples destined for the blood bank [1].

Samples should be collected within 72 h of a scheduled transfusion; otherwise, complement-dependent antibodies may be missed due to instability. New blood samples from recipients are needed for repeat pre-transfusion testing every 72 h if new transfusion orders are made. The age of the samples may be extended to up to 1 month in certain clinical settings, such as preoperative evaluation of elective surgery patients if they have a negative antibody screen and no RBC exposure via pregnancy or blood transfusions within the past 3 months. Blood samples may be rejected; causes for rejection include improper labeling, inadequate amount and hemolyzed sample. If both clerical check (accompanying paperwork) and sample (correct blood collection tube, quantity, etc.) are acceptable, the specimen is processed by immediate centrifugation, and the RBCs and supernatant are separated. Documentation of all steps from receipt of specimen to testing and result interpretation is performed manually or electronically, and records are kept confidential for a period of time in accordance with requirements of federal, state and accrediting agencies (Tables 2.1–2.3).

Table 2.1 Safety of transfusion between various blood groups

Recipient blood group	Donor blood group A: Donor has anti-B antibody	Donor blood group B: Donor has anti-A antibody	Donor blood group AB: Donor has no antibody	Donor blood group O: Donor has anti-A and anti-B antibody
Recipient blood group A	Safe to transfuse		Safe to transfuse	
Recipient blood group B		Safe to transfuse	Safe to transfuse	
Recipient blood group AB			Safe to transfuse	
Recipient blood group O	Safe to transfuse	Safe to transfuse	Safe to transfuse	Safe to transfuse

Table 2.2 Blood component transfusion in a bleeding patient

	PRBCs	FFP	Platelets	Cryoprecipitate
Threshold for transfusion	Hb <8−10 g/dL	INR >1.5	<10,000 without bleeding; <50,000 with bleeding	Fibrinogen <100 mg/dL; uremic thrombocytopathia
Dose (one)	1 unit	2 single units or one jumbo FFP; 10−20 mL/kg	4−6 single units; one apheresis unit	10 units
Expected rise posttransfusion of one dose	Hb by 1 g/dL or Hct by 3%	Unpredictable	30,000−60,000	Increase in fibrinogen by at least 50 mg/dL

Table 2.3 Use of laboratory evidence for transfusion of blood components

Blood component	CBC	DIC screen	TEG
PRBCs	Low Hgb/Hct	NA	NA
FFP	NA	Prolonged PT/PTT. If TT is prolonged with normal fibrinogen levels, heparin may be an issue	TEG R value is prolonged. With heparinase, TEG R value does not decrease
Platelets	Low platelets	NA	TEG MA and angle alpha values are low
Cryoprecipitate	NA	Fibrinogen levels are low	TEG angle alpha value is low

2.2.1 ABO/Rh(D) Typing

Determination of the ABO blood group is performed in two steps, a forward and a reverse reaction. In the forward reaction, the presence of the A and/or B antigens is determined by mixing the RBC to be tested with anti-A and anti-B reagents. If there is reaction with only anti-A reagent, then it can be determined that the patient has the A antigen on the red cells. Therefore, the patient's blood group is A. If there is reaction with anti-B reagent, then the patient has the B antigen and his or her blood group is B. If there is reaction with both anti-A and anti-B reagents, then the patient has both A and B antigens. His or her ABO blood group is AB. If there is no reaction, then the patient has neither A nor B antigens, and his or her blood group is determined as O.

In the reverse reaction, the serum of the patient is mixed with reagent RBCs of A1 and B types to detect the presence of anti–A1 and anti–B antibodies. If there is reaction with A1 cells, then the patient has anti–A1 antibody. If the patient has anti–A1 antibody, then he or she cannot have the A antigen. If there is reaction with B cells, then the patient has anti–B antibody. If the patient has anti–B antibody, then he or she cannot have the B antigen. Thus, an individual with blood group A should have anti–B antibody, whereas an individual with blood group B should have anti–A1 antibody. An individual with blood group AB should show the presence of none of the antibodies, and an individual with blood group O should have both antibodies. If the forward reaction and the reverse reaction lead to different conclusions regarding the ABO type, reaction strength is weaker than expected and/or the historical blood type does not match the current one, the cause for the discrepancy must be fully investigated for a final interpretation of the ABO blood group type.

"Rh typing" is a misnomer because it does not involve phenotyping for all major antigens belonging to the Rh system but, rather, only for the D antigen, which is the most immunogenic out of all known antigens. The Rh or D type is determined in a similar way as the forward ABO typing, using anti–D reagent. Blood donors who type Rh-negative (D-negative) are further tested to detect the presence of so-called weak D antigen (Du) using more sensitive methods, such as the presence of anti-human globulin (AHG) that acts as an enhancer of the reaction between the D antigen and the anti-D reagent.

Determination of weak D is required only for blood donors in order to establish the true D status, and it is performed by testing the patient's RBCs IgG anti-D in the AHG phase. If the weak D testing is positive, the Rh(D) type is interpreted as Rh(D)-positive, and if negative, the Rh(D) type is interpreted as a true negative. The weak D status is neither required nor routinely determined in recipients because Rh(D)-negative blood can be safely transfused regardless of the true Rh(D) status of the recipient. The presence of other antigens belonging to the Rh blood group is not determined for routine transfusions.

2.2.2 Antibody Screen

More than 250 types of antigens are present on the surface of RBCs, and based on their chemical structure, these can be grouped into two major categories—carbohydrates and polypeptides [2]. The RBC antigens are encoded by specific genes and categorized into blood group systems. Major blood group systems include ABO, Rh, Kell, Kidd, Duffy, Lutheran and MNS.

Antigens belonging to the ABO, Lewis, I and P blood group system are carbohydrate in nature. Antibodies that are formed against these antigens do not require sensitization by prior RBC exposure. They are thus called naturally occurring antibodies. They are typically IgM in nature and are considered to be clinically insignificant because they are not associated with hemolytic transfusion reactions or hemolytic disease of the newborn. A major exception to this are the antibodies formed against the antigens of the ABO blood group system. Antibodies to the ABO antigens are naturally occurring but are definitely clinically significant. Antigens belonging to the Rh, Kell, Kidd, Duffy, Lutheran and MNS system are typically IgG in nature, are not naturally occurring and are considered to be clinically significant because they are associated with hemolytic transfusion reactions and hemolytic disease of the newborn. Antibody formation requires prior exposure to RBCs that possess these antigens. Prior exposure may be due to prior transfusion or pregnancy. An exception to this is anti-M and anti-N antibodies, which are typically clinically insignificant.

The antibody screen is performed by mixing the patient's plasma with three reagent RBCs with a known phenotype present in a commercially available kit containing a table of the antigenic profiles of RBC used. This table is called "antigram." For routine pretransfusion testing, the AABB requires that reagent RBC present in the antibody screen should contain at least one homozygous RBC positive for the following major RBC antigens: C, c, D, E, e, Fy^a, Fy^b, Jk^a, Jk^b, K, k, Le^a, Le^b, P_1, M, N, S and s antigens. If the patient's plasma contains alloantibodies against major RBC antigens, the antibody screen becomes positive and further testing by extended panel is required.

The antibody screen and panel are both indirect AHG tests performed with reagent RBCs prepared from donors of type O so that the naturally occurring anti-A or anti-B antibodies do not interfere with the testing. A positive antibody screen implies that recipient plasma might react with antigens present on the donor's red cell membrane. Additional tests are required to determine the specificity of the antibody.

A positive antibody screen can be seen if the patient has underlying alloantibodies or autoantibodies. The blood bank must determine if the underlying antibody is an autoantibody or an alloantibody. If the antibody is an autoantibody, it needs to be determined whether the antibody is a warm autoantibody or a cold antibody. If the underlying antibody is an alloantibody, then the blood bank must determine the specificity of the antibody—that is, the antigen against which the antibody is directed. If the alloantibody is clinically significant, then the blood bank must obtain blood from a donor who lacks that specific antigen to avoid hemolytic transfusion reactions.

2.2.3 Antibody Identification by Extended (Panel) Testing

Positive antibody screens require further testing by testing the patient's plasma against a panel of 10−12 RBCs (commercially available) of varying phenotypes. The identification of antibody specificity is done in two steps. In the first step, RBCs that when tested with the patient's plasma produce no reaction are identified. It may be assumed that if the patient's plasma had antibody specific to the antigens present on these red cells, then there should have been a reaction. Because there is no reaction, the patient's plasma does not have antibodies corresponding to the antigens present on these red cells. Thus, several potential antibodies can be ruled out. In the next step, called "ruling in," the plasma reactivity pattern is compared with the profile of antigens across all cell lines. If at least three cell lines react with the patient's plasma, there is a 95% chance that the reactivity is due to an antibody corresponding to that antigen. Antibodies against major RBC antigens should be ruled out, and any additional reactivity should be explained, if possible. However, often, even if the major alloantibodies are ruled out, an extra weak reactivity might still be present. The clinical significance of this weakly reactive nonspecific (WRNS) antibody is most likely limited if the patient does not have a history of prior RBC exposure, in which case crossmatch–compatible RBC units are provided for the patient. However, if the patient had prior RBC sensitization via pregnancy or recent transfusion, the clinical significance of RBCs is indeterminate, meaning that the developing alloantibody with partial reactivity cannot be completely ruled out. In this case, additional testing, including determination of the patient's RBC phenotype, may be indicated, and the safest blood to be provided for that patient would match the patient's profile.

Occasionally, plasma reacts with all reagent RBCs tested (pan-reactivity); thus, if no negative reactions occur, the process of ruling out antibodies against major RBC antigens cannot be performed. The main differential diagnosis in such cases is autoantibodies, antibody against a high incidence of antigen and multiple antibodies. When plasma reacts with all reagent RBCs and the patient's own RBCs (positive autocontrol), an autoantibody is suspected. Autoantibodies may be of warm type or cold type. When auto-antibodies are present, it may not be possible to rule out underlying alloanti-bodies. In such situations, the patient's RBC phenotype should be determined. This means that the patient's RBCs are tested for the presence of the major antigens. If blood is provided that is matched with the patient's RBC antigens, then even if the patient has any antibody, there should not be any corresponding antigen in the donor blood for it to react with. This is referred to as providing phenotypically matched blood.

2.2.4 Implications of the Presence of Antibodies

Once an antibody is identified, a decision is made regarding whether or not it belongs to a clinically significant class. If it belongs to a clinically significant class, then the patient must receive that particular antigen-negative blood. Some antigens are quite prevalent in the donor popula-tion, and some less so. The blood bank may require additional time to find blood that is negative for a particular antigen. This may be a signifi-cant problem if someone has multiple clinically significant antibodies. If a patient possesses a clinically insignificant antibody, such as a Lewis antibody, the blood bank will not try to find Lewis antigen–negative blood. Rather, it will provide crossmatch–compatible blood.

2.2.5 Crossmatch

Two different types of crossmatch are performed in the blood bank. One is the serologic crossmatch, which tests the blood compatibility between recipient and potential donor. It also serves to reconfirm the ABO type of the donor and thus is a checkpoint for preventing ABO typing errors. There are two types of serologic crossmatch, major and minor. In the major crossmatch, the patient's plasma is tested against donor RBCs obtained from a segment of the blood product. The major crossmatch aims to detect whether the transfused RBCs will react with any antibody in the patient's blood and thus lead to hemolysis. In the minor cross-match, the patient's RBCs are tested against the donor's plasma to search for antibodies in the donor plasma that may react with the patient's red cells; this crossmatch has been discontinued.

The method of performing the major crossmatch depends on the results of the antibody screen. If the antibody screen is negative and the patient does not have a history of clinically significant antibodies, the major crossmatch performed is only in immediate spin phase—so-called incomplete crossmatch. If the antibody screen is positive, the major crossmatch is performed in the AHG phase with selected antigen–negative RBC units. The second method of crossmatch is the "computer crossmatch" or electronic crossmatch, in which the ABO types of both donor and recipient are electronically confirmed and assessment for ABO compatibility by AABB-approved computer software is also acceptable when certain regulatory conditions are met. Chapman et al. commented that the computer crossmatch is a safe alternative to the serological crossmatch [3].

2.3 BLOOD COMPONENTS

Various blood components, including whole blood, are used in transfusion. This section discusses the major blood products and their application in transfusion service.

2.3.1 Whole Blood

Whole blood obtained from donors is typically separated into various components. It is thus rarely used for transfusion directly. Some blood banks still stock units of whole blood for trauma cases. It is typically used for massive blood loss, in the range of 30–40% or more of loss of blood volume. Whole blood is also used for exchange transfusion in neonates. However, in such situations, whole blood is made available by reconstitution by combining RBCs with fresh frozen plasma (FFP). Whole blood may also be used for autologous transfusions.

If used, whole blood must be ABO identical to that of the patient. With storage, levels of labile coagulation factors diminish, as do functional platelets. In general, after collection, whole blood should be cooled to 6°C within 8 h.

2.3.2 Packed Red Blood Cells

Packed red blood cells (PRBCs) are the most commonly used blood component. PRBCs are prepared from whole blood by centrifugation or by apheresis collection. Typically, one unit of PRBC is approximately 350 mL in volume, of which RBC volume is 200–250 mL.

The remaining volume is due to plasma (typically <50 mL), white blood cells (WBCs), platelets and anticoagulants. The most commonly used anticoagulant is CPDA-1 (citrate, phosphate, dextrose and adenine), which allows for 35 days storage at $1-6°C$. The hematocrit of such units is less than 80% (range, 70–80%).

In the United States, most centers provide leukoreduced PRBCs. This is achieved by using leukocyte reduction filters, and such step is performed before storage. Residual leukocytes should not exceed 5×10^6 per unit. Leukocyte-reduced units decrease risk of alloimmunization and reduce the chance of febrile nonhemolytic transfusion reactions. Leukocyte-reduced units are also considered to be cytomegalovirus (CMV) safe and can be given to CMV-negative individuals.

In a nonemergency setting, PRBCs should be transfused at a rate of $1-2$ mL/min for the first 15 min and then increased to 4 mL/min or as rapidly as the patient can tolerate. Transfusion should not exceed 4 h. Potential life-threatening reactions most commonly occur within the first 15 min. In an emergency setting, PRBCs may be transfused at fast rates, and multiple units may need to be transfused. However, transfusion-associated hypothermia, hypocalcemia and hyperkalemia are recognized side effects of such transfusion.

Each unit of PRBC is expected to increase the hematocrit by 3% and the hemoglobin level by 1 g/dL. This effect can be measured 15 min after transfusion. In a bleeding patient, it must be remembered that transfusion of multiple units of PRBC will aggravate the coagulopathic state of the patient. The unit of PRBC that is being transfused must be compatible with the recipient's plasma ABO antibodies. Thus, if the recipient is blood group A, he or she has anti-B antibodies and cannot be transfused with B or AB units. If the recipient is B, he or she has anti-A antibodies and cannot be transfused with A or AB units. Both A and B patients can receive O units. If the recipient is AB, he or she has no anti-A or anti-B antibodies and can therefore receive A, B, AB or O units. If the recipient is O, he or she has anti-A and anti-B antibodies and can receive only O units.

There are various types of PRBCs that may be made available in special circumstances:

- PRBCs leukoreduced: In the United States, most centers provide PRBCs that are leukoreduced. This product deceases the incidence of febrile transfusion reactions. Leukoreduced products are also considered to be CMV safe. Leukoreduction also helps to prevent human leukocyte antigen (HLA) alloimmunization.

- PRBCs irradiated: Irradiated cellular products (RBCs and platelets) are used to prevent transfusion-associated graft-versus-host disease. Examples of indications for irradiation include transfusion to individuals who are immunodeficient, are on intensive chemotherapy, bone marrow transplant recipients and patients with Hodgkin's disease. The maximum storage time after irradiation is 28 days postirradiation or the original expiration time, whichever occurs first. Irradiation damages the red cell membranes, and there is increased leakiness of potassium. When irradiated RBCs are to be given to neonates, the unit needs to be washed to remove the excess potassium.

- PRBCs washed: Washed PRBCs are generally requested when the patient has a history of allergic reactions. Washing is performed with saline, which removes plasma proteins and electrolytes. However, the unit must be transfused within 24 h of washing. Washing of PRBCs (and also platelets) may also be done for transfusion to IgA-deficient patients.

- Frozen RBCs: Sometimes RBCs are frozen to preserve rare donor groups. They can be preserved for up to 10 years. A cryoprotective agent such as glycerol is used to prevent damage to red cells during freezing. When these frozen red cells are thawed for use, the cryoprotective agent (eg, glycerol) must be removed. This is done by washing; thus, these units have to be used within 24 h.

2.3.3 Fresh Frozen Plasma

FFP is prepared from whole blood by separating and freezing the plasma within 8 h of phlebotomy.

The approximate volume of FFP is 200–250 mL. FFP is stored at −18°C or lower for 1 year. Once thawed, it should be used within 24 h. Thawed FFP should be stored at 1–6°C. One milliliter of FFP contains approximately one unit of coagulant factor activity. FFP is most often given to patients with elevated prothrombin time/international normalized ratio (PT/INR). Note that FFP itself has an INR of approximately 1.3 or 1.4. Thus, individuals with an INR of 1.5 or less do not typically need to be transfused with FFP. One conventional dose of FFP in a 70-kg individual is usually considered to be two units of FFP or one jumbo FFP (equivalent to two units of FFP). However, strictly speaking, one dose of FFP is 10–20 mL/kg, and this increases coagulation factors by 20%. One unit of FFP also contains approximately 400 mg of fibrinogen. Thus, mild hypofibrinogenemia will improve with FFP transfusion. Because factor VII has a relatively short half-life ($\sim$ 4 h), the effect of FFP transfusion on INR values may be short-lived.

FFP can be used in the following circumstances:

- Bleeding patients with coagulopathy
- Bleeding patients requiring reversal of warfarin effect
- Correct coagulopathy in anticipation of surgery/invasive procedures
- Exchange fluid for therapeutic plasma exchange (TPE) in certain situations (eg, TPE for thrombotic thrombocytopenia purpura (TTP))
- Part of massive transfusion protocol (RBC:FFP should be 1:1)
- Choice of FFP: FFP has donor antibodies. These antibodies should be compatible with the recipient's red cells

2.3.4 Plasma Variants

There are several plasma variants:

- FP24 is plasma that has been frozen within 24 hr rather than 8 h. In this type of plasma, factor VIII levels are approximately 20—25% less than that of conventional FFP.
- Donor retested (DR) plasma: DR plasma is a unit of donated FFP that has been held until the donor returns at least 112 days later. It has reduced infectivity for HIV-1 and 2, hepatitis C virus (HCV), hepatitis B virus (HBV) and HTLV-1 and 2.
- Solvent detergent-treated (SD) plasma: SD plasma is a pooled plasma product treated with a solvent and detergent to eliminate lipid-enveloped viruses such as HIV-1 and 2, HBV, HCV and HTLV-1 and 2.
- Cryo-poor plasma: This is the residual plasma that remains after cryo-precipitate has been removed. Cryo-poor fraction of FFP has been used in refractory cases of TTP.

2.3.5 Platelets

One unit of platelet may be derived from one unit of donated whole blood by centrifugation. Whole blood has to be centrifuged twice to obtain platelets. The first centrifugation step yields RBCs and platelet-rich plasma. The platelet-rich plasma is spun again to produce platelets and plasma. One unit of platelets has at least 5.5×10^{10} platelets, and the volume is 50 mL. This may be stored at 20—24°C for a maximum of 5 days. Because one typical adult dose is six units, individual units are frequently pooled. If pooled, they should be used within 4 h. Platelets may also be washed, just like RBCs; if washed, they should be used within 4 h.

Platelets may also be obtained by apheresis procedure. One apheresis procedure from one donor will yield at least one dose of platelets. This is equivalent to six units of platelets. The approximate volume is 250–300 mL. The number of platelets in one apheresis platelets is 3.0×10^{11}. The advantage of apheresis platelets is that the recipient is exposed to one donor for one dose of platelet transfusion as opposed to six donors. It is possible to obtain more than one (two or three) dose from one donor by apheresis.

Platelets have platelet-specific antigens, ABO and HLAs. The contaminating RBCs have Rh antigens. However, the amount of contaminating RBCs in one dose of platelets is less than 2 mL of RBCs. Thus, crossmatch is not required for platelets. Whenever possible, however, ABO-compatible platelets should be used. With transfusion of ABO-incompatible units, there arise two important issues. First, the recipient's antibody against the A or B antigens may reduce platelet survival because platelets also have ABO antigens. In reality, in the majority of cases, this does not appear to be significant. In some individuals, this may be an issue when the antigen expression on the platelets is high and the titers of antibodies in the recipients are also high. In such cases, posttransfusion platelet count may be unsatisfactory. This should prompt a trial of group-specific platelet transfusion. Second, platelets are suspended in plasma. The antibodies present from the donor may in theory react with the antigens on the recipient's red cells and cause hemolysis. The risk of hemolysis with apheresis platelets has been shown to be in the range of 1:3000 to 1:10,000. If this risk is to be minimized when choosing ABO-incompatible platelets, a physician should consider the donor platelets as "plasma" and decide on the most suitable units for transfusion.

D-negative individuals should ideally receive platelets from D-negative donors. However, in men and postmenopausal women, this is not an issue. If an Rh-negative premenopausal woman receives Rh-positive platelets, RhIg may be given. Accumulation of cytokines in stored platelets may result in febrile reactions. For this and to prevent alloimmunization and for CMV-negative units, leukocyte-reduced platelets may be used.

Use of platelets:
- Generally indicated for thrombocytopenia and thrombocytopathia
- Not indicated for idiopathic thrombocytopenia purpura, TTP and disseminated intravascular coagulation (DIC)
- Platelet transfusion may also be contraindicated in patients with heparin-induced thrombocytopenia

The usual dose of platelet is 1 unit/10 kg; conventionally, six units or one apheresis dose. A single dose of platelets usually increases the platelet count in a 70-kg person by 30,000–60,000/μL.

Threshold for platelet transfusion:
- For a bleeding patient, the target platelet count is at least 50,000/μL, preferably closer to 100,000/μL. This is especially true for patients with intracerebral, ophthalmic and pulmonary hemorrhage.
- For patients who are not bleeding, prophylactic platelet transfusion may be considered if the count is less than 10,000/μL. This may be increased to 20,000/μL if the patient is coagulopathic, on heparin, or has an anatomic lesion likely to bleed.

Platelet refractoriness:
- When patients receive multiple platelet transfusions, there is a chance that the increase in platelet count will be less than expected. Typically, one dose of platelets should increase the platelet count by 30,000–60,000/μL. Response to platelet transfusion can be determined by measuring platelet counts 10 or 60 min posttransfusion. Calculations are done to obtain a corrected count increment (CCI) value. It is considered that if the CCI is less than 5000 after two consecutive transfusions, then there exists platelet refractoriness.
- Platelet refractoriness is most often due to antibodies against HLA antigens. The recipient is exposed to donor HLA antigens from transfusion of cellular products (RBCs and platelets). This is why leukoreduced products may reduce the incidence of HLA alloimmunization. Platelets possess HLA class I but not class II antigens. Thus, antibodies to HLA class I antigens will result in an inadequate increase in platelet count posttransfusion.
- Other causes of platelet refractoriness include DIC, sepsis, fever, splenomegaly and drugs. In the setting of platelet refractoriness due to HLA antibodies, "HLA-matched" platelets may be transfused. HLA-matched platelets need to be irradiated to prevent graft-versus-host disease.

- Most individuals have the HPA-1a antigen on the surface of their platelets. HPA-1a is also known as PlA1 antigen. If an individual who lacks this antigen receives platelets from a donor who has the antigen, there is a chance of antibody formation against the antigen. This will also potentially cause platelet refractoriness to platelet transfusion. This mechanism as a cause of platelet refractoriness is quite rare. Rather, they are more often associated with neonatal alloimmune thrombocytopenia or posttransfusion purpura.

Formula for CCI:

$$CCI = (Posttransfusion\ platelet\ count) - (Pretransfusion\ platelet\ count)$$
$$\times Body\ surface\ area \times 10^{11}/(Platelets\ transfused \times 10^{11})$$

Similar to PRBCs, platelets can be leukoreduced, irradiated or washed, and these platelets are known as modified platelets. If platelets are obtained from whole blood, then one donor of whole blood will provide one-sixth of a dose of platelets. The six individual aliquots are sometimes pooled prior to transfusion.

2.3.6 Cryoprecipitate

Cryoprecipitate is obtained by thawing frozen plasma at 4°C followed by centrifugation. It may be stored at −18°C or a cooler temperature for up to 1 year. However, once cryoprecipitate is thawed, it should be used within 6 h or within 4 h if pooled. Cryoprecipitate contains the following:

- Fibrinogen (150 mg/unit)
- Factor VIII (80 units/unit)
- Factor XIII (80 units/unit)
- von Willebrand factor
- Fibronectin

Cryoprecipitate is used in treating fibrinogen deficiency, factor XIII deficiency and uremia. The approximate volume of each unit of cryoprecipitate is 5−15 mL. Each unit raises fibrinogen level by 5 mg/dL. The minimum hemostatic level is at least 100 mg/dL. It is prudent, for a bleeding patient, to aim for 200 mg/dL. One conventional dose of cryoprecipitate is 10 units. Note that thawed units or doses of cryo are not maintained in the blood bank. Thus, once ordered, there is a delay of 20−30 min before the product can be made available.

Cryoprecipitate can also be used topically and is applied simultaneously with calcium and bovine thrombin to achieve hemostasis.

However, this may lead to the formation of antibodies to thrombin and other procoagulant proteins in xenogenic products, including factor V. Virus-inactivated fibrinogen concentrates are also available, reducing such uses of cryoprecipitate.

2.4 RELEASE OF BLOOD PRODUCTS

The process of releasing blood products is initiated when the blood bank receives a request for blood products from a clinician as a written form containing all the pertinent information, such as patient information (which includes patient identifiers and red arm band number) and the type and number of blood products required. Blood products are selected from the blood bank inventory. The units are labeled with at least two independent patient identifiers, donor unit number and compatibility test results. The blood components are placed in appropriate transport containers. PRBCs and plasma are placed in coolers, whereas platelets are placed in containers without ice packs. The containers are also labeled with patient particulars, and the containers are handed over or sent to the patient bedside.

Occasionally, blood products must be released on an emergency basis. In such situations, a physician must initiate the request, and a signed form needs to be sent to the blood bank, possibly later. O Rh-negative blood will be sent from the blood bank. The blood bank will continue testing the patient's blood for ABO and RH typing as well as the antibody screen. Once these are done, the patient can then be switched to the appropriate units.

2.4.1 Autologous Blood

Autologous blood may be collected from a patient prior to surgery, at the start of surgery or during the intraoperative period. Use of autologous blood has the advantage of decreasing obvious risks of blood transfusion, such as transmission of infectious diseases and possible transfusion-related complications. Blood can be collected from a prospective patient, typically on a weekly schedule, and then stored for transfusion during surgery. It is assumed that the patient's iron stores are adequate. If multiple donations are required, erythropoietin injection may also be given.

Collection of blood just prior to the start of surgery and then transfused at the end of surgery is known as acute normovolemic hemodilution. Blood that is lost during surgery can be collected under low vacuum pressure into a reservoir. RBCs in such cases are accompanied by activated clotting factors, platelets and cellular debris. Anticoagulant is used to prevent the blood from clotting. When there is sufficient blood in the reservoir, it is pumped into a centrifuge bowl, where it is concentrated and washed with saline. From there, the blood is pumped to an infusion bag. Reinfusion should occur within 4 h after collection to prevent bacterial growth. Typically, approximately half of the blood lost may be salvaged. One typical unit of salvaged blood is 225 mL of saline suspended red cells, with a hematocrit of 50%. Intraoperative blood salvage instruments are referred to as cell savers.

The following are recognized complications of intraoperative blood salvage and subsequent reinfusion:
- Fat and air embolism
- Coagulopathy
- DIC and adult respiratory distress syndrome due to activated platelets and white cells

2.4.2 Transfusion Reactions

Transfusion reactions are hazards of transfusion and are broadly divided into immunologic and nonimmunologic mechanisms. Immunologic transfusion reactions are as follows:
- Febrile nonhemolytic
- Hemolytic: acute and delayed
- Allergic, anaphylactoid and anaphylactic
- Posttransfusion purpura
- Transfusion-related acute lung injury (TRALI)
- Graft-versus-host disease (GVHD)
The following are nonimmunologic transfusion reactions:
- Circulatory overload
- Bacterial contamination/sepsis
- Transmissible infections
- Air embolism
- Hypocalcemia
- Hyperkalemia
- Hypothermia

A common and important manifestation of transfusion reactions is fever. Fever, in the setting of transfusion, is defined as a temperature

elevation of 1°C or 2°F. Fever with or without chills is seen in the following:
- Febrile nonhemolytic transfusion reaction
- Hemolytic transfusion reaction
- Bacterial contamination
- TRALI

Febrile nonhemolytic transfusion reactions are due to cytokines from donor WBCs. WBCs are present in RBCs and platelets. With storage, cytokines accumulate in the bag, and when the unit is transfused this may result in a febrile reaction. With the use of prestorage leukoreduction, the incidence of this particular type of reaction can be significantly decreased. Premedication has no role in the prevention of febrile reactions. There is a 15% chance of recurrence of this type of reaction with future transfusions. If the unit is not leukoreduced, then use of leukofilters at the time of transfusion should help.

Acute hemolytic transfusions are rare and most often are due to clerical errors. The classical features of this type of transfusion reaction include flank pain and hemoglobinuria. Hypotension may also be present and is a useful sign for anesthetized patients to recognize this type of transfusion reaction. Naturally, transfusion should be stopped, the patient should receive fluids, and a potent diuretic should be started to maintain a significant urine output (eg, 100 mL/h or more). Causes of death in such patients are acute tubular necrosis and DIC.

Delayed hemolytic transfusion typically occurs 2—10 days posttransfusion. Alloantibodies to the Kidd and Duffy blood group system are typically implicated in such transfusion reaction. The classical scenario is as follows: The patient lacks one of the antigens of the previously discussed blood group systems. The patient is transfused with red cells with that particular antigen. The patient then forms antibodies against that antigen. Levels of the antibody then decline to below detection levels. Prior to subsequent transfusion, antibody screen is negative and crossmatching does not detect incompatibility. The patient again receives blood that has the same antigen. Antibody levels start to increase, and within 1 week hemolysis occurs. The patient's hemoglobin starts to drop with declining serum haptoglobin. Bilirubin level in serum is elevated. The antibody is coating the surface of the transfused red cells. Thus, the direct antiglobulin test (DAT) will be positive. Fortunately, this type of reaction is not life-threatening. Monitoring the patient and ensuring adequate hydration are effective in most situations.

2.4.3 Transfusion-Related Acute Lung Injury

TRALI is a clinical syndrome that occurs within 6 h of transfusion and is characterized by shortness of breath due to noncardiogenic pulmonary edema, fever and hypotension. TRALI can be seen with any blood products, but most often plasma or platelets are implicated. In the United States, TRALI is currently the leading cause of mortality due to transfusions. The mechanism of TRALI is not fully understood. However, antibodies in the donor product against HLA or neutrophils in the recipient are thought to play a role. Patients develop hypoxemia and may require intubation with mechanical ventilation. Treatment is supportive, and most patients improve with 2—4 days.

2.4.4 Transfusion-Related Graft-Versus-Host Disease

This is a rare but fatal complication and is due to concomitant transfusion of viable lymphocytes from cellular blood products. Normally, the donor lymphocytes are destroyed by the host. However, if the immune system of the recipient is significantly compromised or there is HLA matching of the donor and the recipient, then the donor lymphocytes may not be destroyed. Thus, in situations in which TGVHD is a possibility, the donor products are irradiated to prevent the donor lymphocytes from having the ability to divide.

2.5 TRANSFUSION REACTION WORKUP

Whenever there is a suspicion of a transfusion reaction, the transfusion should be stopped and an intravenous line should be kept open for possible administration of medications and/or fluids. A physician should assess the patient for the possibility of a transfusion reaction; if transfusion reaction is considered, then transfusion reaction workup should be initiated. The blood component is returned to the blood bank.

At the blood bank, a clerical check is performed. The relevant paperwork and the products with labels are checked for clerical error. A sample of the patient's blood is also provided, and a visual inspection for hemolysis is done. The patient's ABO group is rechecked on the posttransfusion sample. A DAT is performed, and the results are compared to those of a pretransfusion one, if available.

If there is a hemolytic transfusion reaction, then visual inspection may reveal hemolysis. The posttransfusion sample DAT may be positive, and this is significant if the pretransfusion DAT is negative. Further testing for haptoglobin levels may be indicated. Haptoglobin levels decline with intravascular hemolysis. Subsequently, serum bilirubin will rise and may also be tested for.

At the conclusion of the transfusion reaction workup, a report issued by the blood bank pathologist will be included in the patient's chart.

2.5.1 RhIg

RhIg is a high-titered anti-D antibody (of human origin) preparation. It is used to prevent alloimmunization to the D antigen. In most instances, RhIg is given to Rh-negative mothers when they are pregnant, if the fathers are Rh-positive. They are also administered to Rh-negative mothers after delivery if they give birth to Rh-positive infants.

In cases of transfusion of red cells and platelets and females who are Rh-negative and of childbearing age, the blood bank will try to provide these products, which are Rh-negative. However, in certain situations, especially with platelet transfusion, this may not be possible. If Rh-positive red cells are given to an Rh-negative individual, use of RhIg is impractical. The patient needs to be tested for the development of anti-D antibody, and if it occurs, this patient must receive Rh-negative blood. Platelets have a small amount of contaminating red cells. In apheresis, donor units (which is one dose of platelets), this typically less than 2 mL. Platelets do not have Rh antigens. However, the contaminating red cells do. Thus, if the platelet product is from an Rh-positive individual and the recipient is Rh-negative, alloimmunization with formation of anti-D antibody is a possibility. To prevent the formation of anti-D antibody in these individuals, RhIg may be given. One dose of RhIg (300 μg) will be sufficient to suppress alloimmunization against 30 mL of whole blood or 15 mL of red cells. Thus, one dose of RhIg is sufficient to prevent alloimmunization of six to eight doses (apheresis donor) of Rh-positive platelets.

2.6 CONCLUSIONS

The blood bank plays an important role in patient care. Transfusion-related errors have serious consequence for patients, including death. Proper identification of the patient and blood products is critical in avoiding such transfusion-related reactions. Most of the errors occur outside the blood bank, so communication between medical staff and blood bank professionals is essential to avoid such errors. In one study, Sharma et al. observed 123 errors in a 10-year period, of which 107 errors (87%) occurred outside the blood bank, whereas 16 errors (13%) occurred within the blood bank [4].

REFERENCES

[1] Goossens W, Van Duppen V, Verwilghen RL. K2- or K3-EDTA: the anticoagulant of choice in routine hematology. Clin Lab Haematol 1991;13:291−5.
[2] Reid ME, Mohandas N. Red blood cell blood group antigens: structure and function. Semin Hematol 2004;41:93−117.
[3] Chapman JF, Milkins C, Voak D. The computer crossmatch: a safe alternative to the serological crossmatch. Transfus Med 2000;10:251−6.
[4] Sharma RR, Kumar S, Agnihotri SK. Souve of preventable errors related to transfusion. Vox Sang 2001;81:37−41.

CHAPTER 3

Pharmacotherapy With Antiplatelet, Anticoagulant, and Their Reversing Agents

Contents

3.1 INTRODUCTION

More than 600,000 adults and 10,000 children undergo cardiac surgery annually in the United States, and the majority of these patients (up to 75%) require transfusion. The estimated annual cost related to transfusion

A. Nguyen, A. Dasgupta, A. Wahed:
Management of Hemostasis and Coagulopathies for Surgical and Critically Ill Patients.   © 2016 Elsevier Inc.
DOI: http://dx.doi.org/10.1016/B978-0-12-803531-3.00003-3   All rights reserved.   **39**

in patients undergoing cardiac surgery is more than $500 million [1]. Both expected and unexpected bleeding occur frequently in patients undergoing cardiac surgery, and one of the goals of patient management is to reduce bleeding because bleeding is often associated with adverse outcomes. Bleeding after surgery can be broadly classified into two categories:

- Bleeding related to surgery
- Inherited disorder of coagulation or acquired coagulopathy

Many factors contribute to bleeding-related complications during and after cardiac surgeries, such as the use of anticoagulants including heparin during cardiopulmonary bypass that may cause multiple hemostatic aberrations, damage to blood vessel walls during the surgical procedure and others. Acquired coagulopathy may be related to hypothermia during surgery, abnormal fibrinolysis or hemodilution or many other factors. Despite reversal of the effects of antifibrinolytic agents using transfusion or protamine, excessive bleeding after cardiac surgery is still a critical issue affecting 3–11% of patients [2]. In general, loss of more than 2 L of blood within the first 24 h after surgery or bleeding at a rate of 300 mL/h or a rate of 100–200 mL/h for 4 h are considered as criteria for excessive bleeding [1].

Antiplatelet and or anticoagulant agents are often used for prevention and treatment of various cardiovascular diseases and also during cardiac surgeries. Common anticoagulants used in these patients include heparin or its derivatives, warfarin and antiplatelet medications such as aspirin (acetylsalicylate), thienopyridine derivatives such as clopidogrel (Plavix) or prasugrel and other agents. The most commonly used antiplatelet agents are aspirin and clopidogrel, although newer drugs are also gaining acceptance among clinicians as excellent antiplatelet agents. Although these agents are effective in preventing thromboembolic complications, a major complication of treating patients with these agents is the risk of hemorrhage, which may be life-threatening. Nevertheless, many clinical studies show a favorable balance between efficacy and safety in using anticoagulants in clinical practice for certain patient populations.

However, prior to cardiac surgery, antiplatelet and anticoagulant agents may be discontinued depending on the clinical picture of the patient. Aspirin (acetylsalicylic acid) is a common medication taken by many patients and is typically discontinued 5–7 days prior to surgery. In cardiac surgery, however, aspirin is typically continued until the day before surgery due to its cardiac protection, despite a low level of bleeding risk. Another common medication is clopidogrel, which must be discontinued for at least 5 days prior to surgery due to significant bleeding risk. Although many patients take warfarin (Coumadin), optimal preoperative warfarin dosing is not firmly established. Bevan reported that a preoperative international normalized ratio (INR) of 1.7 or greater in a patient receiving warfarin is associated with increased bleeding during surgery [3]. In general, warfarin should be stopped at least 3—5 days prior to surgery, and an INR of 1.5 or less is considered safe. After discontinuation of warfarin, these patients may be placed on low-molecular-weight heparin.

However, if bleeding is occurring in a patient requiring emergency cardiac surgery, then reversal of bleeding is needed and various drugs are available for such reversing effects of anticoagulants. For example, vitamin K can be used for reversing the effect of warfarin, protamine can be used for reversing the effect of heparin and platelet can be used for reversing the effect of clopidogrel or aspirin. This chapter provides an overview of various anticoagulants and reversing agents used in managing bleeding of patients undergoing cardiac surgeries. Moreover, clinical laboratories play an important role in managing patients with anticoagulants or during reversing of anticoagulant effects. Laboratory tests necessary to monitor such patients are discussed briefly in this chapter and in more detail in the following chapters.

3.2 ANTIPLATELET AND ANTICOAGULANT AGENTS: AN OVERVIEW

Various antiplatelet and anticoagulant agents are used in clinical practice. The oldest antiplatelet agent is aspirin. Circulating platelets in blood are essential for the formation of blood clots because after platelet aggregation, these aggregates are bound together by fibrin, thus forming blood clots. Antiplatelet drugs prevent platelet aggregation, and based on the mechanism of action, antiplatelet agents can be classified into several categories:

- Cyclooxygenase enzyme inhibitor (aspirin and other nonsteroidal anti-inflammatory drugs)
- Adenosine diphosphate (ADP) receptor blockers (ticlopidine, clopidogrel, prasugrel, ticagrelor and cangrelor)
- Glycoprotein IIb/IIIa inhibitors (tirofiban, abciximab and eptifibatide)
- Phosphodiesterase inhibitors (dipyridamole and cilostazol)
- Thrombin receptor antagonist (vorapaxar)

In addition to antiplatelet drugs, anticoagulants such as warfarin (vitamin K antagonist) and heparin (the antithrombin III−heparin complex is 1000-fold more active than antithrombin III alone) are also widely used in clinical practice. More commonly used drugs in clinical practice are aspirin, warfarin, clopidogrel, abciximab and eptifibatide. For several conditions, such as acute coronary syndrome, unstable angina and non-ST segment elevation myocardial infarction, dual antiplatelet therapy with aspirin and clopidogrel has shown clinical benefits [4].

Thrombin plays a central role in the clotting process because its principal function is to convert soluble fibrinogen into insoluble fibrin and also to stimulate platelet activation. Once formed, thrombin activates factors V, VII and IX, which are involved in generating more thrombins. Thrombin also activates factor XIII, a protein involved in fibrin cross-linking and clot stabilization. Direct thrombin inhibitors are a new class of drugs that directly bind to thrombin and block thrombin's action. These drugs include recombinant hirudins, bivalirudin, argatroban and dabigatran. Another new class of drugs is direct factor Xa inhibitors. These drugs act directly on factor Xa without using antithrombin as a mediator. This class of drugs includes rivaroxaban, apixaban and edoxaban. Pentasaccharides are synthetic compounds that also inhibit factor Xa by selectively binding to antithrombin III (fondaparinux).

However, bleeding is associated with the use of any antiplatelet and/or anticoagulant drug. Therefore, patients should be monitored using appropriate laboratory tests; thus, clinical laboratories play an important role in managing patients receiving such drugs. Moreover, if reversing the effect of such drugs is clinically necessary, appropriate coagulation tests must be performed to ensure that bleeding risk has minimized. Activated partial thromboplastin time (aPTT), thrombin time (TT) and prothrombin time (PT) are effective tests for assessment of coagulation and are frequently used to assess coagulation status of patients receiving an anticoagulant drug or during reversing of the effect of the drug. INR is an excellent indicator of blood clotting. INR is a calculated value based on PT. It is very useful in monitoring therapy with anticoagulants, especially warfarin. VerifyNow (VFN) is a rapid turbidimetric whole blood assay capable of evaluating platelet aggregation. VFN Aspirin assay and VFN Plavix (clopidogrel) assay are useful in monitoring therapy with aspirin or clopidogrel, respectively. Thromboelastography (TEG) is another useful method for assessing global hemostasis and fibrinolytic functions. Platelet mapping is a special TEG assay that measures the effects of antiplatelet drug therapy (aspirin, ADP receptor inhibitors and GpIIb/IIIa inhibitors) on platelet function. The plasma anti–Xa assay may be used for monitoring therapy with unfractionated heparin (UFH) or low–molecular–weight heparin. For an in–depth discussion on coagulation tests, see "Coagulation-Based Tests and Their Interpretation (Chapter 1)."

3.3 REVERSING AGENTS: AN OVERVIEW

Therapy with anticoagulants/antiplatelet agents carries a major risk of overanticoagulation, which results in bleeding episodes that increase the risk of complications and even morbidity and mortality. Warfarin is commonly used in clinical practice as an anticoagulant. Patients experiencing clinically significant bleeding and elevated INR (≥ 4) due to warfarin therapy can be managed by the following:

- Discontinuation of warfarin therapy
- Administration of vitamin K
- Administration of fresh frozen plasma or prothrombin complex concentrate (PCC)

Unfractionated heparin binds to antithrombin III, and the effect of heparin can be reversed by protamine. Protamine can also reverse the effect of low–molecular–weight heparin. For reversing the effect of aspirin, platelet transfusion is indicated. Although platelet transfusion is most effective for rapid reversal of the effect of aspirin, there is evidence in the medical literature for the use of desmopressin (deamino-D-arginine vasopressin) for reversing the effect of aspirin. In our institution, we have not used desmopressin for this purpose.

Platelet transfusion is also effective in reversing the effects of ADP receptor blockers. Whereas vitamin K, platelet, fresh frozen plasma and protamine are the main reversal agents for correcting the effects of anticoagulants, there is no established reversal strategy for thrombin inhibitors, although the half–life of these agents is short and discontinuation of therapy may be sufficient for reversing their effects. For reversing the effects of factor Xa inhibitors, discontinuation of therapy may also be sufficient. Hemodialysis is effective in removing dabigatran due to its low protein binding.

Recombinant factor VIIa was introduced in the 1980s and is an excellent agent for the prevention and treatment of severe bleeding. This agent is used primarily in treating patients with congenital or acquired hemophilia, but currently it has many other uses. Recombinant factor VIIa is effective in reversing the effects of synthetic polysaccharides such as fondaparinux.

Another available reversible agent is desmopressin, an arginiue analogue of vasopressin that increases plasma factor VIII and von Willebrand factor. Although used primarily in treating mild von Willebrand disease and hemophilia A, tranexamic acid and amino-caproic acid are synthetic lysine analogues that exert an antifibrinolytic effect by attaching to the lysine binding site of plasminogen, thus displacing plasminogen from fibrin. Both drugs are effective in reducing blood loss 30–40% during cardiac surgery. However, tranexamic acid is significantly more potent than aminocaproic acid but also has a longer half-life.

3.4 ASPIRIN

Aspirin is rapidly absorbed after oral administration from the stomach and intestine by passive diffusion, and oral bioavailability is approximately 70%. Aspirin has a short half-life in plasma ($\sim$ 30 min) and is hydrolyzed by enzyme esterase present in both gastrointestinal mucosa and the liver into active metabolite salicylate. Although salicylate is responsible for the anti-inflammatory and analgesic effect of aspirin, the antiplatelet effect is mainly due to the parent drug aspirin (acetylsalicylate). Salicylate is strongly bound to serum protein ($\sim$ 82%), mostly to serum albumin.

The primary antiplatelet action of aspirin is due to irreversible inhibition of cyclooxygenase enzyme (COX). This enzyme exists in two isoforms, COX-1 and COX-2, and converts arachidonic acid into prostaglandin, which is then converted into several bioactive prostanoids, including thromboxane A_2 (TXA_2) and prostacyclin (PGI_2). Aspirin blocks catalytic sites of both COX-1 and COX-2 by binding to the active sites and acetylating a serine residue (serine residue at position 529 in human COX-1 and serine residue at position 516 in human COX-2). This prevents arachidonic acid from gaining access to the catalytic site [5]. Aspirin can be dosed once a day because platelet inhibition is permanent and the effect of aspirin can only be reversed by generation of new platelets. Dosage of aspirin for antiplatelet effect varies from 75 to 325 mg; however, under certain clinical conditions, higher dosage may also be used.

The major adverse effect of aspirin is increased risk of bleeding complications, and the most common site for bleeding is the gastrointestinal tract. Low-dose aspirin is effective in preventing cardiovascular diseases in men, and use of low-dose aspirin as a prophylactic agent has been suggested for men older than age 45 years with risk factors for cardiovascular diseases. The benefit of aspirin in low dosage (75–162 mg per day) in prevention of myocardial infarction in men and thrombolytic stroke in women outweighs the risk of bleeding [6]. Low-dose aspirin may also decrease the risk of cancer. In the United States, low-dose aspirin is available as chewable or enteric coded tablets at doses of 81 or 162 mg.

Aspirin resistance is defined by an inability of aspirin to inhibit COX-1-dependent TXA_2 production. Current estimates suggest that up to 30% of treated individuals may have inadequate response to aspirin at dosage less than 300 mg per day. Other causes of aspirin resistance include polymorphism of platelet glycoprotein receptor, COX-1 as well as COX-2 alleles; generation of aspirin-insensitive COX; and higher platelet turnover. Higher aspirin dosage or use of another antiplatelet drug, such as clopidogrel alone or in combination with aspirin, may be appropriate to overcome aspirin resistance in some patients [7]. Many other nonsteroidal anti-inflammatory drugs (NSAIDs), such as ibuprofen, naproxen, tolmetin, indomethacin, diclofenac, ketorolac and diflunisal, also inhibit COX enzymes but are mostly used as analgesic as well as anti-inflammatory drugs. These agents are rarely used for their antiplatelet activities. Various pharmacokinetic features of aspirin are summarized in Table 3.1.

3.4.1 Monitoring Aspirin Therapy

For monitoring aspirin therapy, appropriate platelet function testing should be performed. Currently, point-of-care testing such as VerifyNow Aspirin, Plateletworks and PFA-100 are available for monitoring aspirin therapy. The VerifyNow Aspirin test is a rapid turbidimetric whole blood assay that evaluates platelet aggregation of fibrinogen-coated beads in response to arachidonic acid that results in optical changes. These changes are

Table 3.1 Summary of pharmacokinetic parameters of aspirin

- Aspirin (acetyl salicylate) is rapidly absorbed after oral administration with approximate bioavailability of 70%.
- Dosage of aspirin may vary from 75 to 325 mg or more.
- Usually low-dose aspirin is defined as 75−162 mg of aspirin per day.
- Aspirin has a short half-life in blood (~30 min) and is hydrolyzed by enzyme esterase present in both gastrointestinal mucosa and the liver into active metabolite salicylate.
- Salicylate is responsible for anti-inflammatory and analgesic effects of aspirin.
- Salicylate is strongly bound to serum protein (~82%) mostly to serum albumin.
- Only the parent drug aspirin has antiplatelet effect.

converted into aspirin response units (ARUs), and the manufacturer suggests that a value less than 550 ARUs is consistent with adequate aspirin response. Plateletworks is a platelet function test that uses whole blood and estimates platelet function by comparing the platelet count before and after exposure with arachidonic acid. The functional platelets should aggregate, but nonfunctional platelets (aspirin's action) should not aggregate. A hematology analyzer (impedance cell counter: ICHOR II analyzer) is used to count platelets both before and after adding arachidonic acid. Because platelet aggregates are not counted by the analyzer, platelet count should be less after adding arachidonic acid. The manufacturer suggests that individuals taking aspirin should have less than 60% aggregation. The PFA-100 system is a platelet function analyzer designed to measure platelet-related hemostasis. The PFA-100 analyzer evaluates platelet aggregation in high shear condition in response to agonist cartridge containing collagen and epinephrine. The endpoint is closure of small aperture due to platelet aggregation. Aspirin should prolong the closure time.

There is also a biochemical test to monitor aspirin therapy. The TXA_2 has a short half-life and is converted into stable inactive metabolites TXB_2 and 11-dehydro-TXB_2, which are excreted in urine. An enzyme-linked immunosorbent assay (AspirinWorks, Corgenix, Broomfield, CO) can measure 11-dehydro-TXB_2 in a random urine sample. Creatinine in the urine must also be measured. The manufacturer suggests that adequate response to aspirin is reflected by a value of 1500 pg of 11-dehydro-TXB_2 per milligram of creatinine [8]. Management of acute aspirin-induced hemorrhage requires platelet transfusion to increase the platelet count by 50,000/μL. Desmopressin may also be helpful. Because aspirin has an irreversible effect on platelets, the coagulopathy may last for 4 or 5 days after discontinuation of aspirin therapy, and platelet transfusions may have to be repeated daily.

3.5 ADENOSINE DIPHOSPHATE RECEPTOR BLOCKERS

ADP is an important mediator of platelet aggregation. The effect of ADP on platelets is mediated by two P2Y receptors (a family of purinergic G protein-coupled receptors): PGY_1 and PGY_{12}. Activation of the G_q pathway through the PGY_1 receptor leads to platelet shape change and rapidly reversible platelet aggregation, whereas activation of the G_i pathway through PGY_{12} activation amplifies G_q-mediated response, resulting in sustainable platelet aggregation. PGY_{12} activation may also activate glycoprotein IIB/IIIa integrin through phosphoinositide (PI) 3-kinase pathways and other G protein-based pathways, thus playing a vital role in the irreversible wave of platelet aggregation [9]. Several antiplatelet drugs, such as ticlopidine, clopidogrel, prasugrel and ticagrelor, exert their antiplatelet effects through PGY_{12} receptor blocking. These drugs can be used with aspirin or alone for antiplatelet therapy.

3.5.1 Ticlopidine and Clopidogrel

Ticlopidine is the first-generation oral thienopyridine that exerts its antiplatelet activity by inhibiting ADP-induced platelet aggregation (PGY_{12} receptor blocker). The active metabolite of ticlopidine is responsible for irreversible PGY_{12} receptor blocking. Bioavailability is approximately 80−90%, and the half-life is 12 or 13 h. The recommended loading dose is 500 mg, and then the drug should be administered twice daily at a dosage of 250 mg. However, use of ticlopidine is limited due to its bone marrow toxicity (neutropenia); as a result, second-generation oral thienopyridine drugs such as clopidogrel were developed. Clopidogrel is now a standard therapy across the spectrum of patients with coronary artery diseases and also patients undergoing various cardiac surgeries [10]. Currently, use of ticlopidine has been discontinued in the United States.

Clopidogrel, which also belongs to the thienopyridine class of organic compounds, is more effective than aspirin for the prevention of vascular disease without increasing bleeding complications compared to use of aspirin. Clopidogrel can be used in patients who are allergic to aspirin. Clopidogrel can be administered alone or in combination with aspirin for increased protection from thrombosis; however, when clopidogrel is combined with aspirin, bleeding risk is also increased. Like aspirin, clopidogrel can be administered orally; after administration, clopidogrel is rapidly absorbed with approximately 50% bioavailability. However, clopidogrel is a prodrug with no antiplatelet activity. Therefore, clopidogrel must be converted into an active metabolite that is pharmacologically active. This active metabolite is formed rapidly by actions of liver enzymes CYP3A4/CYP3A5 through intermediate formation of 2-oxo-clopidogrel. However, antiplatelet activity of clopidogrel may be affected by functional polymorphism of the *CYP3A5* gene-encoding CYP3A5 enzyme. Other cytochrome P450 enzymes, such as CYP1A2, CYP2B6, CYP2C9 and CYP2C19, are also involved in some aspects of metabolism of clopidogrel. In a competitive pathway, approximately 85% of the drug is hydrolyzed by liver esterase into an inactive metabolite (carboxylic acid derivative).

The antiplatelet effect of clopidogrel is due to the active metabolite, which contains a thiol group that binds to the free cysteine residue on the PGY_{12} receptor (forming disulfide bridges between cysteine residue position Cys17 and Cys270), thus irreversibly blocking ADP binding and receptor activation. The recommended loading dosage varies from 300 to 600 mg, but then the drug can be administered orally once per day at 75-mg dosage. The pharmacological activity lasts for approximately 7−10 days, the entire life span of platelets. The half-life of active metabolite is approximately 30 min. Studies have indicated that 20−40% of patients may not respond to clopidogrel. Disease states such as old age, diabetes, renal failure and cardiac failure may cause lower than expected pharmacological action of clopidogrel. Genetic polymorphism of the PGY_{12} receptor, genetic polymorphism of the *CYP2C19* gene (the presence of nonfunctional alleles such as *CYP2C19*2, -*3, -*4*, or *-*5*) or alteration of the intracellular signaling mechanism may also be associated with lower or nonresponse to clopidogrel [11].

3.5.2 Prasugrel

Prasugrel is a third-generation thienopyridine class of organic compounds and acts by its active metabolite, which irreversibly inhibits the PGY_{12} receptor by forming disulfide bridges between cysteine residues (position Cys17 and Cys270), thus preventing platelet aggregation. In the clinical setting, prasugrel produces a greater degree of platelet inhibition than clopidogrel and is associated with fewer cardiac events, such as myocardial infarction, recurrent ischemia and clinical target vessel thrombosis, than clopidogrel. Similar to clopidogrel, prasugrel is also a prodrug that must be converted into active metabolite mostly by CYP3A4 and, to a lesser extent, CYP2B6. Compared to clopidogrel, conversion of prasugrel into active metabolite is more rapid, and this is the reason why prasugrel is more potent than clopidogrel as an antiplatelet agent. Moreover, genetic polymorphism of CYP2C19 and/or CYP2C9 has minimal effect on metabolism of prasugrel compared to clopidogrel.

After oral administration, prasugrel is rapidly absorbed with an approximate bioavailability of 80%. The drug is strongly bound to albumin (98%). Peak metabolite concentration is observed in approximately 30 min. Approximately 68% of dose is excreted as metabolites in the urine, and the rest is excreted in feces. Initial recommended loading dose is 40−60 mg, and daily maintenance dose is 10−15 mg. In general, the maximum antiplatelet effect of prasugrel is observed 2 days after initiation of therapy, and after discontinuation of the drug, platelet function is gradually recovered within 2 days. Compared to clopidogrel, there are few nonresponders to prasugrel. However, due to higher potency of prasugrel compared to clopidogrel, the risk of bleeding is also higher with prasugrel therapy compared to clopidogrel [7].

3.5.3 Ticagrelor

Currently, the standard of treatment for acute coronary syndrome is dual antiplatelet therapy consisting of aspirin and a PGY_{12} receptor blocker such as clopidogrel or prasugrel. However, a major limitation of clopidogrel and prasugrel is their irreversible PGY_{12} receptor binding, which results in a prolonged time for recovery of platelet function after discontinuation of therapy. In addition, both drugs are prodrugs, and metabolic activation is needed for their pharmacological actions. Ticagrelor is the first oral reversible direct-acting inhibitor of the PGY_{12} receptor that does not require any metabolic activation for its pharmacological action. Chemically, ticagrelor belongs to the cyclopentyltriazolopyrimidine class, and it is a high-affinity nucleoside analogue of ADP.

After oral administration, ticagrelor is rapidly absorbed and is metabolized to at least one active metabolite with similar pharmacokinetic parameters to the parent drug. Peak plasma concentration as well as maximum inhibition of platelet aggregation are observed 1−3 h after oral administration. The mean plasma half-life is 6−12 h. The loading dose is usually 180 mg, and maintenance dosage is 90 mg administered twice a day. The superiority of ticagrelor compared to clopidogrel was demonstrated in the PLATO (Platelet Inhibition and Patient Outcome) study involving 18,624 patients hospitalized due to acute coronary syndrome. Ticagrelor treatment was associated with a significant reduction in the number of deaths due to vascular causes, myocardial infarction or stroke compared to clopidogrel. In addition, major bleeding episodes were similar between patients receiving ticagrelor (11.6%) and those receiving clopidogrel (11.2%). Overall, ticagrelor therapy is associated with faster, greater and more comprehensive consistent inhibition of platelet aggregation than clopidogrel. However, therapy with ticagrelor may be associated with increased nonprocedure-related bleeding, dyspnea, and ventricular pauses in the first week of treatment [12]. Various pharmacokinetic features of ticlopidine, clopidogrel, prasugrel and ticagrelor are summarized in Table 3.2.

Table 3.2 Summary of pharmacokinetic parameters of PGY_{12} receptor blocking agents

Drug name (trade name)	Summary of pharmacokinetic parameters
Ticlopidine (Ticlid)	• Ticlopidine, a first-generation oral thienopyridine, exerts its antiplatelet activity by inhibiting adenosine diphosphate-induced platelet aggregation (PGY_{12} receptor blocker). • Ticlopidine is a prodrug, and the active metabolite is responsible for irreversible PGY_{12} receptor blocking. • Bioavailability is approximately 80–90%. • Half-life is 12 or 13 h. • The recommended loading dose is 500 mg, and then the drug should be administered twice daily at a dosage of 250 mg. • Use of ticlopidine is limited due to its bone marrow toxicity (neutropenia); as a result, second-generation oral thienopyridine drugs such as clopidogrel were developed.
Clopidogrel (Plavix)	• Clopidogrel is a second-generation oral thienopyridine that exerts its antiplatelet activity by inhibiting adenosine diphosphate-induced platelet aggregation (PGY_{12} receptor blocker). • Clopidogrel is a prodrug, and the active metabolite is responsible for irreversible PGY_{12} receptor blocking. • Bioavailability is approximately 50%. • Half-life is 30 min for the metabolite. • The recommended loading dose is 300–600 mg, and then the drug should be administered once daily (75 mg dose). • Genetic polymorphism of CYP2C9 may affect metabolism.
Prasugrel (Effient)	• Prasugrel is a third-generation oral thienopyridine that exerts its antiplatelet activity by inhibiting adenosine diphosphate-induced platelet aggregation (PGY_{12} receptor blocker). • Prasugrel is a prodrug, and the active metabolite is responsible for irreversible PGY_{12} receptor blocking. • Bioavailability is approximately 80%, and the drug is 98% bound to serum protein. • Half-life of the parent drug is approximately 7 h. • The recommended loading dose is 40–60 mg, and then the drug should be administered once daily (10–15 mg). • It is a superior antiplatelet drug compared to clopidogrel.

(*Continued*)

Table 3.2 (Continued)

Drug name (trade name)	Summary of pharmacokinetic parameters
Ticagrelor (Brilinta)	• Ticagrelor is a cyclopentyltriazolopyrimidine class of drug that is a high-affinity nucleoside analogue of ADP. • Parent drug is active, and it is reversibly bound to PGY_{12} receptor. • There is at least one active metabolite. • Bioavailability is approximately 36%, and the drug is greater than 99% bound to serum proteins. • Half-life is 6−12 h for the parent drug. • The recommended loading dose is 180 mg, and then the drug should be administered twice daily (90 mg dose).

3.5.4 Cangrelor

Cangrelor is a reversible high-affinity PGY_{12} receptor blocker that was relatively recently approved by the U.S. Food and Drug Administration (FDA) for antiplatelet therapy. This drug is administered intravenously and has a rapid onset of action; platelet inhibition starts within 2 min and lasts for the duration of infusion. Cangrelor does not require metabolic activation for its pharmacological effect. The steady state is achieved within 30 min (in the absence of loading dose), and the drug has a short half-life of less than 9 min. The effect of this drug is quickly reversed after discontinuation as platelet function is rapidly returned within 60 min of discontinuation of therapy.

Another drug, elinogrel, which has a novel sulfonylurea structure, is also a direct-acting reversible PGY_{12} receptor blocker. It may be administered both intravenously and orally, but currently further development of this drug has been discontinued.

3.5.4.1 Monitoring Therapy with ADP Receptor-Blocking Agents

Monitoring platelet function is the best way to monitor therapy with ADP receptor-blocking agents. Clopidogrel, like aspirin, prolongs closure time of the aperture. VerifyNow PGY_{12} assay can be used to monitor therapy with clopidogrel. In this assay, adenosine-5-diphosphate (ADP/PGE1) is incorporated into the test channel to induce platelet activation without fibrin formation. The values are represented as Plavix reactive units (PRUs), and values less than 210 represent adequate response to clopidogrel (Plavix). TEG can demonstrate alteration in platelet aggregation but cannot demonstrate ADP blockage caused by clopidogrel. There is no specific reversal agent for therapy with clopidogrel or related drugs. Transfusion with normal platelet may be used for reversal of effects.

3.6 GLYCOPROTEIN IIB/IIIA INHIBITORS

The glycoprotein IIb/IIIa inhibitors, such as abciximab, eptifibatide and tirofiban, are currently available and must be administered intravenously. During clotting, platelets are activated due to binding of a variety of agonists (platelet-activating factor, ADP, epinephrine, serotonin, vasopressin, collagen, etc.) on different receptors present on the platelet cell surface. Once bound to damaged tissue, platelets are activated and undergo conformational changes at glycoprotein IIa/IIIb receptor sites, which allows binding of fibrinogen and von Willebrand factor, causing platelet aggregation. The glycoprotein IIb/IIIa inhibitors interfere with platelet activity during the final stage, thus preventing platelet-induced thrombosis. In addition to preventing platelet aggregation, these drugs can also induce dissolution of platelet-rich clot by disrupting fibrogen—platelet interaction. The glycoprotein IIb/IIIa inhibitors are widely used for patients undergoing percutaneous coronary intervention (PCI) including angioplasty, atherectomy and stent placement. The FDA has also approved these drugs for treatment of acute coronary syndrome, including unstable angina and non-Q-wave myocardial infarction [13].

Abciximab (Reopro) is a large monoclonal antibody (molecular weight, 47,600 Da) that binds noncompetitively but with high affinity to glycoprotein IIb/IIIa receptor. This drug is intended for patients undergoing PCI for preventing serious heart problems, including myocardial infarction. It is also used in certain patients with unstable angina to prevent serious heart problems when PCI is planned. Abciximab is used in combination with other medicines, such as heparin and aspirin. The plasma half-life is short (10—30 min). Dissociation from glycoprotein IIb/IIIa receptor occurs through proteolysis, which is a slow process. Therefore, the antiplatelet effect of the drug lasts approximately 6—12 h. Detectable platelet inhibitions may be observed for up to 2 weeks. An initial bolus dose of 0.25 μg/kg is recommended prior to intervention. A continuous infusion of 0.125 μg/kg/min is administered over 12 h. Maximum administered dose is 10 μg/min. Abciximab should be discontinued 12—24 h before procedure. Currently, there is no available reversal agent. Therefore, platelet transfusion is recommended for reversing the effect of abciximab.

Eptifibatide (Integrilin) is approved for use in patients with acute coronary syndrome, including patients scheduled for PCI. Eptifibatide consists of low-molecular cyclic peptide (molecular weight, 832 Da), which has higher binding specificity but lower binding affinity with glycoprotein IIb/IIIa receptors compared to abciximab. This drug has a longer plasma half-life (~ 2.5 h) compared to abciximab. Kidneys account for approximately 60−70% clearance of the drug; as a result, dosage adjustment is needed for patients with impaired kidney function. However, dosing adjustment may not be needed for patients with hepatic impairment. Prior to PCI, two subsequent boluses of 180 µg/kg are administered in 10 min, followed by a continuous infusion of 2 µg/kg/min for 18−24 h. Eptifibatide should be discontinued 2−4 h before any procedure. There is no available reversible agent for eptifibatide. Therefore, platelet transfusion is the option to reverse the effect of the drug.

Tirofiban (Aggrastat) is a low-molecular-weight nonpeptide (molecular weight, 495 Da) that competitively inhibits glycoprotein IIb/IIIa receptor. In contrast to eptifibatide, which is a peptide, tirofiban is a low-molecular-weight nonpeptide molecule. The plasma half-life is approximately 1.6 h, and tirofiban undergoes both renal and nonrenal clearance, with renal clearance accounting for approximately 65% clearance of the drug. As a result, dosage reduction is needed for patients with renal insufficiency. An initial bolus dose of 10 µg/kg is administered in 3 min, followed by infusion of 0.15 µg/kg/min for 18−24 h. However, a bolus dose of 25 µg/kg has also been proposed. Tirofiban can be stopped before the procedure, but currently there is no available reversible agent. As a result, platelet transfusion may be needed for reversing the effect. Various clinical parameters of abciximab, eptifibatide and tirofiban are summarized in Table 3.3.

3.7 PHOSPHODIESTERASE INHIBITORS

Phosphodiesterase inhibitors such as dipyridamole and cilostazol are used in antiplatelet therapy. Dipyridamole (Persantine) is a pyrimidopyrimidine derivative that inhibits the enzyme phosphodiesterase in platelets, thus increasing intraplatelet levels of cyclic adenosine monophosphate (cAMP) and cyclic guanosine monophosphate (cGMP), which eventually inhibit platelet aggregation. The increased cAMP and cGMP inside platelets also

Table 3.3 Various clinical parameters of abciximab, eptifibatide and tirofiban

Parameter	Abciximab	Eptifibatide	Tirofiban
Administration	Intravenously	Intravenously	Intravenously
Dosage	Continuous infusion of 0.125 µg/kg/min for 12 h. Maximum administered dose is 10 µg/min	Two subsequent boluses of 180 µg/kg are administered in 10 min, followed by a continuous infusion of 2 µg/kg/min for 18–24 h	An initial bolus dose of 10 µg/kg is administered in 3 min followed by infusion of 0.15 µg/kg/min for 18–24 h
Onset of action	Immediate	Immediate	Immediate
Plasma half-life	20–30 min	2.5 h	1.6 h
Duration of antiplatelet effect	6–12 h	2–4 h	2–4 h
Metabolism	Unbound drug via proteolytic cleavage	Approximately 60–70% renal clearance; therefore, dose reduction needed for renally impaired patients	Approximately 65% renal clearance; dose reduction needed for renally impaired patients
When the drug should be discontinued before procedure	12–24 h	2–4 h	Can be discontinued just before procedure
Reversal agent	None; platelet transfusion (one dose) is recommended for reversing the effect of abciximab	None; platelet transfusion (one dose) is recommended for reversing the effect of abciximab	None; platelet transfusion (one dose) is recommended for reversing the effect of abciximab

potentiate platelet inhibitory effect of prostacyclin. Dipyridamole also inhibits cellular uptake and metabolism of adenosine, which further contributes to inhibition of platelet aggregation. Although dipyridamole alone is not superior to other platelet drugs and its side effect profile, mainly headache, limits its application, dipyridamole in combination with low-dose aspirin (aspirin 25 mg with extended-release dipyridamole 200 mg; Aggrenox) given twice daily orally is effective in preventing stroke and transient ischemic attack. Dipyridamole is strongly bound to serum protein, and the extended-release formulation has a plasma half-life of 13 h. The drug is metabolized to a glucuronide and excreted mainly in bile. Dipyridamole is also subjected to enterohepatic recirculation [14].

Cilostazol (Pletal) is also a phosphodiesterase inhibitor that inhibits platelet aggregation but also has vasodilatory property. Cilostazol is primarily indicated for treatment of intermittent claudication, and this drug helps patients to walk longer and faster before developing pain. Cilostazol has been shown to prevent stent thrombosis and restenosis. The effect of cilostazol is enhanced if used as a part of triple therapy along with aspirin and clopidogrel. Side effects are mostly headache, skin rash and stomach upset. The drug is given orally in a dosage of 50 or 100 mg twice a day and should be administered 30 min before a meal or 2 h after a meal to avoid food–drug interaction. Bioavailability of cilostazol is variable. The drug is primarily metabolized by CYP3A4/5 and, to a lesser extent, CYP2C19 to inactive metabolites. The elimination half-life is approximately 10 h, and no dosage adjustment is needed for patients with impaired renal or liver function [7].

3.8 THROMBIN RECEPTOR ANTAGONIST

In the setting of non-ST segment elevated acute coronary syndrome, optimal management includes antiplatelet therapy with multiple drugs. Aspirin irreversibly blocks COX-1 and impedes formation of TXA_2 that results in decreased platelet activation. As a result, aggregation of platelet is inhibited by aspirin. Direct inhibition of thrombin-mediated platelet aggregation may be an alternative mechanism of action of an antiplatelet agent. Thrombin proteolytically activates cell surface protease activated receptors (PARs; four such receptors have been identified— PAR-1, 2, 3 and 4), also known as thrombin receptors. Vorapaxar (Zontivity) is a synthetic analogue of the natural product himbacine, which is a highly active competitive inhibitor of PAR-1 (thrombin receptor) and thus an inhibitor of platelet aggregation. This antiplatelet agent can be administered orally. The loading dosage may be 5, 10, 20 or 40 mg, with a maintenance dosage of 0.5, 1.0 or 2.5 mg given daily preferably in the morning. Voraparax is effective in the secondary prevention of cardiovascular events in stable patients with peripheral artery disease or a history of myocardial infarction along with aspirin and/or clopidogrel.

After oral administration, vorapaxar is absorbed from the gastrointestinal tract, and peak antiplatelet effect is observed 1 or 2 h after oral administration. The drug can be administered with food. Vorapaxar is metabolized by CYP3A4 and is mainly excreted in feces (renal clearance <5%). The elimination half-life is very long (159—311 h) [15].

3.9 WARFARIN

The coumarin-type anticoagulants, such as warfarin, phenprocoumon and acenocoumarol, have been used as anticoagulants for a long time. Warfarin (Coumadin) is the most commonly used drug in this category, commonly prescribed for treatment of venous and arterial thromboembolic disorders. Warfarin is a synthetic compound first developed by the Wisconsin Alumni Research Foundation in 1947. The name "warfarin" derives from WARF (Wisconsin Alumni Research Foundation) plus "arin," indicating its link to the natural product coumarin. Warfarin was initially marketed in 1948 as a rodent poison, but later its usefulness as an anticoagulant for humans was discovered and this drug has been used in clinical practice since the 1950s.

Currently, warfarin, a water-soluble drug, is available as a racemic mixture of 50% R-warfarin and 50% S-warfarin sodium salt. The S-warfarin is four times more potent than R-warfarin. Vitamin K is essential for activation of various clotting factors (II, VII, IX and X), and in this process, vitamin K is oxidized to vitamin K epoxide. Through the action of enzyme vitamin K epoxide reductase complex 1 (VKORC1), vitamin K epoxide is converted back into vitamin K in the liver. Warfarin and its related compounds act as vitamin K antagonists by inhibiting VKORC1; as a result, the hepatic synthesis of various blood clotting factors, such as prothrombin (factor II), factor VII, factor IX and factor X, is impaired. In addition, warfarin interferes with the action of other anticoagulant proteins, such as protein C and protein S [16].

Following initiation of warfarin therapy, 5—7 days are needed for its anticoagulant effect because preexisting clotting factors must be degraded naturally before defective clotting factors are produced by the liver after starting warfarin therapy. Because warfarin therapy has an induction period, patients with preformed clots—for example, patients with pulmonary embolism or deep vein thrombosis—may need a short-term therapy with heparin. The three most common indications for warfarin therapy are atrial fibrillation, venous thromboembolism and prosthetic heart valves. Warfarin is strongly bound to serum albumin (97−99%), and it is the small free fraction of warfarin (free warfarin) that exerts its pharmacological action. The elimination half-life of warfarin varies greatly between 35 and 45 h, but the S-isomer has an average half-life shorter than that of the R-isomer.

Warfarin dosing is affected by the genetic polymorphism of the *CYP2C9* gene encoding the CYP2C9 isoform of the liver cytochrome P450 mixed function oxidase family of enzymes as well as vitamin K epoxide reductase complex subunit 1 gene (*VKORC1*). Moreover, warfarin can interact with many drugs and food, and such interactions can be either pharmacokinetic or pharmacodynamic in nature. Given all these factors that may affect the effectiveness and safety of warfarin therapy, pharmacodynamic monitoring of warfarin therapy is critical. The most commonly adopted way to monitor warfarin therapy is maintaining an acceptable prothrombin time expressed as the INR. However, INR reflects a complex physiologic endpoint of warfarin therapy that is affected by both vitamin K-dependent (II, VI, and X) and independent (I and V) clotting factors. An INR value of 1.0 is the average value for healthy individuals not on warfarin or other anticoagulant therapy. The target INR value for patients receiving warfarin is often 2.5, although the INR target value can range from 1.5 to 3 [17]. The risk of hemorrhage increases at INR values greater than 4.0. Depending on the patient, it may take weeks to even 1 month to reach a stable INR. Flexible protocols, computer programs and nomograms have helped to improve control, but there are still significant limitations in these approaches. Nevertheless, the major risk of bleeding from warfarin therapy has been reported to range from 1% to 3% per year [18].

When initiating warfarin therapy, clinicians should avoid loading doses that can raise INR excessively; instead, warfarin should be initiated at a dose of 5 mg (or 2−4 mg for elderly patients). Sconce et al. reported that incorporation of age, *CYP2C9* and *VKORC1* genotype and height of the patient allowed the best warfarin maintenance dose [19]. The common polymorphism of the *CYP2C9* gene includes the *CYP2C9*2* isoform, which is due to replacement of arginine at amino acid residue 144 by cysteine. This mutation reduces the catalytic activity of the enzyme to approximately 12% of the wild-type enzyme. The CYP2C9*3 allele is due to substitution of leucine for isoleucine at amino acid position 359, which results in a significant reduction of catalytic activity of the enzyme to approximately 5% of wild-type. Therefore, as expected, both CYP2C9*2 and CYP2C9*3 genetic variants, which are common in Caucasian populations, with allele frequencies varying from 3.3% to 18%, result in significantly impaired metabolism of warfarin compared to the wild-type CYP2C9 enzyme [20]. Thus, these individuals require much lower dosage of warfarin as well as a longer time to achieve stable warfarin dosing, and they are at high risk of bleeding from warfarin therapy [21].

Moyer et al. reported that polymorphism in CYP2C9 and VKORC1 genes has considerable effects on warfarin dosing among Caucasian patients [22]. Caucasian individuals show considerable variability in CYP2C9 allele types, whereas people of Asian and African descent infrequently carry inactive CYP2C9 allelic variants. The VKORC1AA allele associated with high warfarin sensitivity predominates in Asians, whereas people of Caucasian and African descent have VKORC1BB (low warfarin sensitivity), VKORC1AB (moderate warfarin sensitivity) or VKORC1AA allele associated with high warfarin sensitivity [22]. In light of the effect of polymorphism of CYP2C9 and VKORC1 genes on warfarin dosage, the FDA changed the label of warfarin, suggesting that clinicians consider genetic testing before initiation of warfarin therapy.

There are many significant interactions between warfarin and various drugs. Such interactions may potentiate or reduce efficacy of warfarin. Both pharmacokinetic and pharmacodynamic interactions between warfarin and drugs have been reported. Some antibiotics interact pharmacokinetically with warfarin by interfering with hepatic metabolism of warfarin, whereas other antibiotics may inhibit bacterial flora in the intestine, thus reducing the amount of vitamin K produced by intestinal bacteria. Many drugs with antiplatelet effects also interact with warfarin pharmacodynamically. The most common drugs in this category are aspirin and NSAIDs. Even selective serotonin reuptake inhibitors used for treating depression have some antiplatelet effect and may interact with warfarin. Therefore, in patients in whom there is a change in preexisting medications, closer monitoring of INR (within 3–7 days of change of medication) is strongly recommended.

Although aspirin potentiates the effect of warfarin, low-dose aspirin (75—100 mg) can be used along with warfarin in treating patients as a part of anticoagulation therapy. Interaction between a vaccine and warfarin is uncommon, but there is one report of fatal intracranial bleeding potentially due to interaction of warfarin with influenza vaccine. Interaction of warfarin with various herbal supplements is of major clinical significance. Although interaction of warfarin with St. John's wort (induces liver enzymes and reduces the effectiveness of warfarin) and milk thistle (inhibits liver enzymes and increases the effectiveness of warfarin) is pharmacokinetic in nature, many clinically significant interactions of warfarin with herbal supplements are pharmacodynamic in nature. In addition, some herbal supplements contain coumarin, which may exert an additive effect with warfarin, causing excessive anticoagulation. Major drug—drug interactions involving warfarin are listed in Table 3.4. Major drug—herb interactions are listed in Table 3.5.

3.9.1 Reversal of Warfarin Therapy

As mentioned previously, INR monitoring is essential in patients receiving warfarin to minimize risk of bleeding. A prolonged INR without clinically evident bleeding may necessitate discontinuation of warfarin therapy. Vitamin K is an effective reversal agent for warfarin. However, vitamin K may be combined with fresh frozen plasma or PCC for faster reversal of the effect of warfarin. PCC is pooled plasma that contains factors II, IX and X, with variable factor VII, protein C and protein S. Refer to the chapter "Antiplatelets and Anticoagulants" for more details on reversal of warfarin.

3.10 HEPARIN AND LOW-MOLECULAR-WEIGHT HEPARIN

Heparin is a natural occurring mixture of various polysaccharides with different molecular weights that is present in human and animal tissues, most commonly liver and lungs. Heparin was originally isolated from canine liver, and the name was derived from the Greek word *hepar*, meaning liver. Commercially available heparin is isolated from a bovine or porcine source, and crude heparin requires purification before use. Unfractionated heparin is a heterogeneous mixture of polysaccharides with molecular weight varying from 5000 to 30,000 Da, and the number of saccharide units in heparin molecules varies from 5 to 35. Heparin itself has no anticoagulant effect, and it binds with antithrombin III

Table 3.4 Examples of clinically important interactions of warfarin with common drugs[a]

Interacting drug	Effect/mechanism
Drugs potentiating the effect of warfarin	
Antibiotics [ciprofloxacin, clarithromycin, erythromycin, metronidazole, bactrim (sulfamethoxazole/trimethoprim)]	Increased effect of warfarin (INR) due to inhibition of intestinal flora producing less vitamin K
Acetaminophen	Increased INR and increased risk of bleeding
NSAIDs	Increased INR and increased risk of bleeding due to antiplatelet activity of NSAIDs; direct mucosal injury by NSAIDs may also increase risk of bleeding during therapy with warfarin
Low-dose aspirin (75–100 mg)	May be used as a therapy in combination with warfarin
Amiodarone	Increased INR and risk of bleeding because amiodarone inhibits CYP2C9
Clofibrate	Potentiation of effect of warfarin by affecting coagulation factors
Antifungal (fluconazole, miconazole, voriconazole)	Increased INR and increased risk of bleeding due to inhibition of CYP2C9
Antidepressants (SSRIs, Fluoxetine, fluvoxamine, etc.)	Increased INR due to some antiplatelet activity of SSRI; fluoxetine also may inhibit CYP2C9
Cimetidine, omeprazole	Potentiation of effect due to stereoselective clearance of R-isomer
Clopidogrel, prasugrel	Increased effect of warfarin and increased risk of bleeding due to antiplatelet effect of these drugs
Ticagrelor	Increased effect of warfarin and increased risk of bleeding due to antiplatelet effect of ticagrelor
Drugs inhibiting the effect of warfarin	
Rifampin	Decreased INR and decreased effect of warfarin because rifampin induces CYP2C9
Carbamazepine	Decreased effect of warfarin because carbamazepine induces liver enzymes
Barbiturates, chlordiazepoxide	Decreased effect of warfarin due to induction of enzymes

INR, international normalized ratio; NSAIDs, nonsteroidal anti-inflammatory drugs; SSRIs, selective serotonin reuptake inhibitors.
[a]These are examples of drug interaction involving warfarin. This list does not include all drugs because more than 800 drugs are known to interact with warfarin.

Table 3.5 Interaction of warfarin with dietary supplements and vegetables

Interacting herbs	Dietary supplement	
Herbs potentiating the effect of warfarin		
Vegetables rich in vitamin K that potentiates effect of warfarin	Angelica root	Bromelain
	Chamomile	Dong quai
	Devil's claw	Fenugreek
	Feverfew	Grape seed
	Garlic	Ginkgo biloba
	Fish oil supplement	Horse chestnut
	Saw palmetto	Danshen
	Spinach	Brussels sprout
	Asparagus	Cabbage
	Kale	Turnip green
	Broccoli	Parsley
	Okra	
Herbs reducing the effect of warfarin		
	St. John's wort	Milk thistle
	Goldenseal	Ginseng

(ATIII), an endogenous inhibitor of thrombin whose physiological role is to prevent coagulation by blocking the activity of thrombin. A minimum of 18 saccharides are needed for binding of heparin to ATIII. However, inhibition of factor Xa by heparin molecules may occur even if the molecule has fewer than 18 saccharides, but a critical pentasaccharide sequence is needed for such effect. Heparin molecules lacking such critical sequence may possess alternative anticoagulant activity.

When a heparin molecule containing 18 or more polysaccharides binds with ATIII, this complex becomes 1000-fold more potent than ATIII itself in preventing coagulation by blocking thrombin and its enzymatic conversion of fibrinogen into fibrin that provides a solid matrix of the clot. Heparin also inhibits clotting by interfering with other procoagulant effects of thrombin, such as thrombin-induced activation of clotting factors V and VIII, which accelerates clot formation. The heparin—ATIII complex also enhances the neutralizing effects of ATIII on activated coagulation factors IX, X, XII and XIII, as well as kallikrein, which contributes to clot formation. In addition, heparin binds to heparin cofactor II, a glycoprotein that inactivates thrombin independent of ATIII. Heparin must be administered intravenously or by subcutaneous injection. It is very effective in stopping clot formation in arteries of patients with cardiovascular diseases; during cardiac surgery; preventing clot formation in patients undergoing hemodialysis or hemofiltration; treating deep vein thrombosis; during orthopedic surgery and neurosurgery; and treating trauma patients [23].

Pharmacokinetic parameters of UFH are variable. UFH is extensively bound to low-density lipoprotein, globulins and also fibrinogen. The plasma half-life is 30—90 min in healthy subjects but increases with increasing dosage. The drug is mainly removed from circulation by the reticuloendothelial system, and enzymes such as heparinase and desulfatase inactivate heparin. UFH is also eliminated by the renal route, and patients with renal impairment may show a longer half-life for heparin. Dose and route of administration depend on the clinical situation being treated with heparin. For prevention of venous thrombosis, heparin is administered subcutaneously in the abdominal fat layer with a typical dose of 5000 units every 8—12 h. Dosage should be adjusted by measuring aPTT. However, a weight-based dosing scheme is superior for achieving desired therapeutic response from heparin. For weight-based protocol, 80 units/kg is proposed as a loading dose (maximum 1000 units) and a maintenance dose of 18 units/kg/h not exceeding 2300 units/h. If aPTT is greater than 90 s, heparin infusion must be stopped.

Apart from major bleeding, which affects approximately 2% of patients, heparin-induced thrombocytopenia (HIT), which affects 1—5% of patients, is the most frequent complication of heparin therapy. Although the exact mechanism of HIT is unclear, research suggests that heparin forms a complex with platelet factor 4 (PF4) and other proteins released by platelets. This complex triggers the production of IgG antibodies directed against itself, causing an autoimmune response. HIT may be mild or severe.

3.10.1 Low-Molecular-Weight Heparin

Low-molecular-weight heparins are prepared from unfractionated heparin by either chemical depolymerization or heparinase digestion (tinzaparin). The molecular weights vary from 4400 to 6500 Da. The low-molecular-weight heparins are parenterally administered drugs and include ardeparin (average molecular weight, 5500—65,000 Da), bemiparin (average molecular weight, 3600 Da), certoparin (average molecular weight, 5400 Da), dalteparin (average molecular weight, 6000 Da), enoxaparin (average molecular weight, 4500 Da), nadroparin (average molecular weight, 4300 Da), parnaparin (average molecular weight, 5000 Da), reviparin (average molecular weight, 4400 Da) and tinzaparin (average molecular weight, 6500 Da). Compared to unfractionated heparin, low-molecular-weight heparins show a more predictable dose—response curve. Low-molecular-weight heparins can be administered as fixed dosage based on body weight, and they reach peak level 2—4 h after subcutaneous administration. Average half-life is 3 or 4 h, and approximately 80% of drugs are eliminated via the renal route. As a result, patients with renal insufficiency require dosage adjustment. Antifactor Xa level can be used to monitor therapy with low-molecular-weight heparin; ideally, this should be measured 4 h after administration of low-molecular-weight heparin.

3.10.2 Reversible Agents for Heparin

UFH therapy should be monitoring by measuring aPTT, whereas therapy with low-molecular-weight heparin should be monitored by using anti-factor Xa assay. The pharmacological effect of both UFH and low-molecular-weight heparin can be reversed by using protamine sulfate. In general, 1 mg of protamine sulfate given intravenously will neutralize 100 units of heparin given in the previous 4 h. Protamine neutralizes all anti-thrombin effects of low-molecular-weight heparins but incompletely reverses factor Xa inhibition. In general, 1 mg of protamine given intravenously will neutralize 1 mg of enoxaparin or 100 units of dalteparin, as well as 100 units of tinzaparin given within 8 h. A second dosage of 0.5 mg protamine may be administered if bleeding continues.

3.11 DIRECT THROMBIN INHIBITORS

Thrombin plays a key role in the generation of thrombus because it converts soluble fibrinogen into insoluble thrombin and also activates various factors as well as factor XII, which is involved in fibrin cross-linking and clot formation. Direct thrombin inhibitors bind directly to thrombin and, unlike heparin, do not require a cofactor such as antithrombin III to exert their pharmacological action. Most direct thrombin inhibitors are administered parenterally (lepirudin, desirudin, bivalirudin and argatroban), but dabigatran can be administered orally. These drugs are used as both prophylaxis and treatment of acute coronary syndrome and venous thromboembolism. These drugs are also used for preventing thromboembolic complications in patients with HIT or in those at risk of HIT and undergoing PCI.

3.11.1 Parenterally Administered Direct Thrombin Inhibitors

Lepirudin (Refludan) and desirudin (Iprivask) are derivatives of hirudin, a peptide originally isolated from the salivary glands of medicinal leeches. Lepirudin is used for treating thrombosis–related complications due to HIT and is given as intravenous infusion with or without bolus. Recommended dosage is 0.4 mg/kg bolus followed by 0.15 mg/kg/h infusion. Because lepirudin is excreted renally, dose adjustment is needed for patients with impaired renal functions. One major complication of treatment with lepirudin is the formation of antihirudin antibody in approximately 40% of patients following treatment of HIT. These immunogenic complexes delay renal excretion of lepirudin, thus increasing its toxicity due to accumulation of the drug. As a result, dose adjustment using aPTT monitoring is required. Lepirudin also has a narrow therapeutic window and potential for increased bleeding risk. Currently, no reversal agent is available for lepirudin.

Desirudin is administered subcutaneously for preventing venous thromboembolism postoperatively in patients undergoing elective hip replacement surgery. Dosage is subcutaneous injection 15 mg twice daily. After subcutaneous administration, maximum plasma level is reached after 1–3 h. The plasma half-life is 2 h. The majority of drug (80–90%) is excreted renally, and for patients with renal impairment (creatinine clearance 30 mL/min or less), dosage adjustment is necessary. For renally impaired patients, aPTT monitoring is required, but aPTT monitoring or dosage adjustment may not be needed for patients with moderate renal insufficiency.

Bivalirudin (Angiomax) is an engineered 20-amino acid synthetic analogue of hirudin with a weaker thrombin inhibitory activity than that of hirudin. Moreover, the binding of bivalirudin to thrombin is reversible. As a result, bleeding risk is minimized. Bivalirudin is effective in reducing death, events of myocardial infarction or repeat vascularization in patients with acute coronary syndrome undergoing PCI. Bivalirudin is also a safe anticoagulant during PCI procedure in patients with HIT. Studies indicate that bivalirudin is associated with lower bleeding risk compared to unfractionated heparin. Bivalirudin can also be used along with glycoprotein IIb/IIIa inhibitors in patients undergoing PCI. Bivalirudin is administered intravenously with dosage of bolus 0.75 mg/kg followed by 1.75 mg/kg/h infusion for 4 h. The drug has immediate onset of action, as evidenced by observing therapeutically desirable activated clotting time

(ACT) within 5 min. The plasma half-life is approximately 25 min. The drug is eliminated by proteolytic cleavage and hepatic metabolism. Although 20% of the dose is renally excreted, dosage adjustment is needed even in patients with moderate renal insufficiency, and drug is contraindicated in patients with severe renal impairment.

Argatroban (Acova) is a small molecule (molecular weight, 527 Da) that noncovalently as well as reversibly binds to the active site of thrombin. This drug is used as a prophylaxis or treatment of thrombosis in patients with HIT as well as patients with a history of HIT or risk of HIT who are undergoing PCI. This drug is administered intravenously with a dose of 2 μg/kg/min. An initial bolus dose is not required. The recommended dosage is 2 μg/kg/h. The plasma half-life is approximately 45 min. The drug is excreted through the biliary system; as a result, dose adjustment is needed in patients with liver diseases but not in patients with impaired renal failure. Therapy can be monitored using aPTT or ACT results. When used with warfarin, INR is further prolonged [24]. Currently, there is no available reversal agent for parentally administered direct thrombin inhibitors. Packed red blood cells or fresh frozen plasma may be administered to counteract the effect.

3.11.2 Orally Administered Direct Thrombin Inhibitors

Dabigatran is a small molecule (molecular weight, 472 Da) that directly inhibits thrombin by binding to its active site via ionic interactions. Dabigatran rapidly and reversibly inhibits both clot-bound and free thrombin. However, due to poor bioavailability (6% or 7%), dabigatran is administered orally as dabigatran etexilate (Pradaxa), a lipophilic and gastrointestinally absorbed substance that after absorption is converted into dabigatran through two intermediates where serine esterases is responsible for cleavage of ester bonds (dabigatran etexilate is a double prodrug).

Efficacy of dabigatran in preventing venous thromboembolism after hip or knee replacement surgery has been well established. It is also effective in secondary prevention of coronary events after acute coronary syndrome, as well as in preventing thrombosis in patients undergoing PCI. After oral administration of dabigatran, bioconversion into active dabigatran occurs in enterocytes, hepatocytes and the portal vein. The peak plasma concentration is observed at 1½−2 h, and the plasma half-life is 12−14 h. Steady-state concentration is achieved in 3 days. Renal excretion is the primary route of elimination of dabigatran. A part of the drugs is excreted in the bile after conjugation with glucuronic acid. Dosage adjustment is required for patients with renal impairment. TT is more responsive to dabigatran than aPTT or PT. However, the ecarin clotting time test is most appropriate for monitoring therapy with dabigatran [24]. Currently, there is no available reversal agent for parentally administered direct thrombin inhibitors. Packed red blood cells or fresh frozen plasma may be administered to counteract the effect. For patients with impaired renal function who are experiencing a life-threatening bleeding episode due to dabigatran–induced effect, some experts recommend hemodialysis. Various clinical parameters of direct thrombin inhibitors are summarized in Table 3.6.

3.12 FACTOR XA INHIBITORS

As expected, factor Xa inhibitors inhibit factor Xa, the first step in the common pathway for coagulation, in a dose-dependent manner. Rivaroxaban, apixaban and edoxaban directly bind to the active site of factor Xa, thus inhibiting both free and clot–associated factor Xa. Indirect inhibitor such as fondaparinux binds to ATIII, causing conformational change that results in the inhibition of factor Xa without having any effect on factor IIa.

Rivaroxaban (Xarelto) is an orally administered reversible factor Xa inhibitor that selectively blocks the active site of factor Xa. Rivaroxaban is an oxazolidinone derivate that has structural similarity with antibiotic linezolid, but rivaroxaban has no antibacterial activity. Rivaroxaban is the first factor Xa inhibitor that was approved by the FDA in 2011 and is used as a prophylaxis to prevent venous thromboembolism in patients undergoing total hip or knee replacement surgery. Later, the FDA also approved rivaroxaban for use as a prophylactic stroke-preventing agent in patients with nonvalvular atrial fibrillation. In large clinical trials, oral rivaroxaban 10 mg once daily showed more effectiveness than subcutaneously administered once-daily 40-mg enoxaparin (a low–molecular–weight heparin) in preventing venous thromboembolism in patients undergoing total hip or knee replacement surgery because venous thromboembolism occurred in 1.1% patients receiving rivaroxaban but 3.7% of patients receiving enoxaparin. Moreover, although rivaroxaban showed greater efficacy than enoxaparin, it did not increase the bleeding risk, the major side effect of anticoagulants, compared to enoxaparin [25].

Table 3.6 Various clinical parameters of direct thrombin inhibitors

Parameter	Lepirudin	Desirudin	Bivalirudin	Argatroban	Dabigatran etexilate
Chemical composition	Recombinant hirudin	Recombinant hirudin	20-Amino acid synthetic analogue of hirudin	Small molecule (527 Da)	Small molecule (427 Da) administered as prodrug dabigatran etexilate
Dosage	Bolus 0.4 mg/kg followed by 0.15 mg/kg/h	Subcutaneous injection 15 mg twice daily	Bolus 0.75 mg/kg followed by 1.75 mg/kg/h for 4 h	No bolus but infusion rate of 2 μg/kg/h	150 mg twice daily; if creatinine clearance between 15 and 30 mL/min, then 75 mg twice daily
Route of administration	Intravenous	Subcutaneous/ intravenous	Intravenous	Intravenous	Oral
Thrombin affinity	High	High	Intermediate	Low	Intermediate to low
Plasma half-life	80 min	Subcutaneous, 120 min; intravenous, 60 min	25 min	45 min	12–14 h
Monitoring	aPTT	Not needed	ACT after 5 min	aPTT	ECT or TT
Metabolism	Kidney	Kidney	Liver and 20% by kidney	Liver	Kidney
Dosage reduction	Renal impairment	Renal impairment	Renal impairment and contraindicated with creatinine clearance <15 mL/min	Liver disease and contraindicated in hepatic failure	Renal impairment

ACT, activated clotting time; aPPT, activated partial thromboplastin time; ECT, ecarin clotting time; TT, thrombin time.

At a dose of 10 mg, bioavailability of rivaroxaban is 80—100%. However, bioavailability is dose dependent because with a 20-mg oral dose, bioavailability is approximately 66%. Peak plasma concentration is observed 2—4 h after administration. Recommended initial dose is 15 mg twice a day for the first 21 days, followed by maintenance dose of 20 mg once a day with evening meal in patients with normal renal function (creatinine clearance >50 mL/min). For patients with severe renal impairment, dosage should be reduced. Rivaroxaban should not be used in patients with Child—Pugh B or C hepatitis impairment. The drug is highly bound to plasma proteins (90—95%), where albumin is the major binding protein. One-third of the drug is excreted unchanged by the kidneys, whereas drug metabolites are excreted both in urine and in feces. Liver enzymes such as CYP3A4/5 play an important role in metabolism of rivaroxaban, accounting for metabolism of approximately 51% of the drug. The plasma half-life of the drug is 5—9 h, but it may be longer in the elderly. Therapy with rivaroxaban is not usually monitored, although it prolongs both PT and aPTT. However, antifactor Xa assay may be used if monitoring is intended. Currently, no reversal agent is available, but recombinant factor VIII and PCC may be administered to reverse the effect of rivaroxaban.

Apixaban (Eliquis) is another orally administered drug that is a reversible factor Xa inhibitor approved for the prevention of venous thromboembolism in patients after total hip or knee replacement surgery. This drug is also used for reduction in the risk of stroke or systematic embolism in patients with nonvalvular atrial fibrillation. In addition, it may be used in the treatment of venous thromboembolism. Apixaban has approximately 50% bioavailability, and drug absorption is not affected by food. For patients with nonvalvular atrial fibrillation, the recommended dosage is 5 mg twice daily, but dosage may be reduced to 2.5 mg in elderly patients (>80 years of age), in those with lower body weight (≤60 kg), or in those with evidence of renal impairment (serum creatinine >1.5 mg/dL). The plasma protein binding is 87%, and plasma half-life is 12 h. Apixaban is metabolized mostly through CYP3A4, and renal clearance accounts for approximately 27% excretion of the drug. Apixaban is also excreted in feces, accounting for approximately 56% of recovered dosage. In a clinical trial involving patients with atrial fibrillation, apixaban (5 mg twice daily) was superior to warfarin in preventing systematic embolism. Moreover, major bleeding episodes occur more frequently in patients receiving warfarin than apixaban [26].

Edoxaban (Savaysa) is the most recently approved (January 2015) oral factor Xa inhibitor indicated for reducing risk of stroke or systematic embolism in patients with nonvalvular atrial fibrillation. It may also be used in treating deep vein thrombosis and pulmonary embolism after initial therapy using a parenterally administered anticoagulant. For patients with nonvalvular atrial fibrillation, recommended dosage is 60 mg once daily. However, edoxaban should be avoided in patients with nonvalvular atrial fibrillation who have creatinine clearance greater than 95 mL/min because the drug may not be effective in such patients with near normal or normal kidney function. This drug has a quick onset of action and peak plasma concentration is achieved within 1 or 2 h after oral administration. The bioavailability is approximately 60%, and the drug can be taken with food. The mean elimination half-life is estimated to be 10−14 h, and steady state is reached within 72 h. Approximately 50% of the drug is excreted unchanged by the kidneys, whereas another 50% is excreted via the biliary and intestinal route. Cytochrome P450 enzymes are minimally involved in the metabolism of edoxaban, accounting for only 4% of metabolism, but the drug is metabolized via hydrolysis. Human carboxylesterase-1 forms M4, a major metabolite of edoxaban that is pharmacologically active [27]. Various parameters of rivaroxaban, apixaban and edoxaban are summarized in Table 3.7.

Fondaparinux (Arixtra; molecular weight, 1728 Da) is a synthetic methoxy derivative of the natural pentasaccharide (five monomeric sugars) sequence, which is identical to the sequence present in heparin that is responsible for high-affinity binding with ATIII. However, unlike low-molecular-weight heparin, fondaparinux is specific for indirect inhibition of factor Xa through inhibiting ATIII and has no effect on thrombin or any other clotting factors. Fondaparinux is administered subcutaneously, and dosage is 2.5 mg/day for prevention of venous thromboembolism in major orthopedic surgery (hip fracture or total hip/knee replacement). The initial dosage should be administered 6−8 h after surgery. In patients with acute symptomatic deep vein thrombosis or pulmonary embolism, recommended dosage is 5 mg (body weight <50 kg) or 7.5 mg (body weight between 50 and 100 kg). The maximum dosage is 10 mg per day for body weight greater than 100 kg. The maximum plasma concentration is observed in 2 h, and elimination half-life is approximately 17 h, which allows for once-daily dosage. The drug is metabolized unchanged through the kidneys; as a result, dose reduction is needed for patients with impaired kidney functions. Fondaparinux is

Table 3.7 Various clinical parameters of rivaroxaban, apixaban and edoxaban

Parameter	Rivaroxaban	Apixaban	Edoxaban
Administration	Oral	Oral	Oral
Dosage	Initial dose: 15 mg twice a day for first 21 days followed by maintenance dose of 20 mg once a day with evening meal	5 mg twice daily but may be reduced to 2.5 mg in elderly patients (>80 years of age), patients with lower body weight (≤60 kg), or patients with evidence of renal impairment (serum creatinine >1.5 mg/dL)	60 mg once daily
Bioavailability	66–100%	50%	60%
Protein binding	90–95%	87%	90–95%
Plasma half-life	5–9 h but may be longer in elderly	12 h	10–14 h
Metabolism	One-third of the drug is excreted unchanged by the kidneys while drug metabolites are excreted both in urine and in feces. CYP3A4/5 enzymes account for 51% of metabolism.	Metabolized mostly through CYP3A4 with 27% renal clearance. Approximately 56% of dosage is recovered in feces.	Endoxaban is metabolized by hydrolysis, and CYP3A4/5 metabolizes only 4% of the drug. Carboxylesterase converts edoxaban to an active metabolite M4.
Reversal agent	No known reversal agent, so administration of recombinant factor VIII/PCC is recommended	No known reversal agent, so administration of recombinant factor VIII/PCC is recommended	No known reversal agent, so administration of recombinant factor VIII/PCC is recommended

contraindicated in patients with severe renal insufficiency (creatinine clearance <30 mL/min). However, for patients with mild to moderate hepatic impairment, no dosage adjustment is needed [28]. Recombinant factor VIIa is effective in reversing the effect of fondaparinux.

3.13 CONCLUSIONS

Many antiplatelet and anticoagulant agents are available for pharmacotherapy for use as prophylaxis in patients undergoing PCI or orthopedic surgeries involving hip or knee replacement. Traditionally used drugs, such as aspirin, warfarin and heparin (including low-molecular-weight heparin), are also widely used clinically today despite the availability of newer agents. Selection of the proper drug depends on the clinical situation and preference of the physician. However, for reversing the effect of these reagents, fresh frozen plasma, platelet transfusion or transfusion with PCC are major options, although protamine can reverse the effect of unfractionated heparin and low-molecular-weight heparin. Moreover, vitamin K can reverse the effect of warfarin. Clinical monitoring of these patients using INR, ACT, aPTT or other appropriate coagulation tests is also necessary for certain agents.

REFERENCES

[1] Steiner ME, Despotis GJ. Transfusion algorithms and how they apply to blood conservation: the high risk cardiac patients. Hematol Oncol Clin North Am 2007;21:177—84.
[2] Despotis GJ, Frilos KS, Zoys TN, et al. Factors associated with excessive postoperative blood loss and hemostatic transfusion requirements: multivariant analysis in cardiac surgical patients. Anesth Analog 1996;82:13—21.
[3] Bevan DH. Cardiac bypass hemostasis: putting blood through the mill. Br J Haematol 1999;104:208—19.
[4] Yusuf S, Zhao F, Mehta SR, et al. Effects of clopidogrel in addition to aspirin in patients with acute coronary syndromes without ST-segment elevation. N Engl J Med 2001;345:494—502.
[5] Catella-Lawson F, Reilly MP, Kapoor SC, Cucchiara AJ, et al. Cyclooxygenase inhibitors and the antiplatelet effect of aspirin. N Engl J Med 2001;345:1809—17.
[6] Colwell JA. Aspirin for the primary prevention of cardiovascular events. Timely Top Med Cardiovasc Dis 2006;10:E25.
[7] Hall R, Mazer CD. Antiplatelet drugs: a review of their pharmacology and management in the perioperative period. Anesth Analg 2012;112:292—318.
[8] Smock KJ, Rodgers GM. Laboratory evaluation of aspirin responsiveness. Am J Hematol 2010;85:358—60.
[9] Fontana P, Dupont A, Gandrille S, Bachelot-Loza C, et al. Adenosine diphosphate induced platelet aggregation is associated with $P2Y_{12}$ gene sequence variation in healthy subjects. Circulation 2003;108:989—95.

[10] Chang JW. Updates in antiplatelet agents used in cardiovascular diseases. J Cardiovasc Pharmacol Ther 2013;18:514—24.

[11] Sangkuhl K, Klein TE, Altman RB. Clopidogrel pathway. Pharmacogenet Genomics 2010;20:463—5.

[12] Kowalczyk M, Banach M, Mikhailidis DP, Hannam S, et al. Ticagrelor—a new platelet aggregation inhibitor in patients with acute coronary syndromes. An improvement of other inhibitors? Med Sci Monit 2009;15:MS24—30.

[13] Stangl PA, Lewis S. Review of currently available GP IIb/IIIA inhibitors and their role in peripheral vascular interventions. Semin Intervent Radiol 2010;27:412—21.

[14] Lenz T, Hilleman DE. Aggrenox: a fixed dose combination of aspirin and dipyridamole. Ann Pharamcother 2000;34:1283—90.

[15] Abdulsattar Y, Taenas T, Garcia D. Vorapaxar: targeting a novel antiplatelet pathway. PT (Pharmacy and Therapeutics) 2011;36:564—8.

[16] Haustein KO. Pharmacokinetic and pharmacodynamic properties of oral anticoagulants, especially phenprocoumon. Semin Thromb Hemost 1999;25:5—11.

[17] Pengo V, Cucchini U, Denas G, Davidson BL, et al. Lower versus standard intensity oral anticoagulant therapy (OAT) in elderly warfarin-experienced patients with nonvalvular atrial fibrillation. Thromb Haemost 2010;103:442—9.

[18] Lee A, Crowther M. Practical issues with vitamin K antagonists: elevated INRs, low time in therapeutic range, and warfarin failure. J Thromb Thrombolysis 2011;31:249—58.

[19] Sconce EA, Khan TI, Wynne HA, Avery P, et al. The impact of CYP2C9 and VKORC1 genetic polymorphism and patient characteristics upon warfarin dose requirements: proposal for a new dosing regimen. Blood 2005;106:2329—33.

[20] Ross KA, Bigham AW, Edwards M, Gozdzik A, et al. Worldwide allele frequency distribution of four polymorphisms associated with warfarin dose requirements. J Hum Genet 2010;55:582—9.

[21] Caraco Y, Blotnick S, Muszkat M. CYP2C9 genotype-guided warfarin prescribing enhances the efficacy and safety of anticoagulation: a prospective randomized controlled study. Clin Pharmacol Ther 2008;83:460—70.

[22] Moyer TP, O'Kane DJ, Baudhuin LM, Wiley CL, et al. Warfarin sensitivity genotyping: a review of the literature and summary of patient experience. Mayo Clin Proc 2009;84:1079—94.

[23] Hassell K. Anticoagulation in heparin induced thrombocytopenia: an ongoing challenge. Hospital Physician 2000;36:56—61.

[24] Lee CJ, Ansell JE. Direct thrombin inhibitors. Br J Clin Pharmacol 2011;72:581—92.

[25] Duggan ST. Rivaroxaban: a review of its use for the prophylaxis of venous thromboembolism after total hip or knee replacement surgery. Am J Cardiovasc Drugs 2012;12:57—72.

[26] McCarty D, Robinson A. Factor Xa inhibitors, a novel therapeutic class for the treatment of nonvalvular atrial fibrillation. Ther Adv Cardiovasc Dis 2016;10(1):37—49.

[27] Zalpour A, Oo TH. Update on edoxaban for the prevention and treatment of thromboembolism: clinical applications based on current evidence. Adv Hematol 2015;2015:920361.

[28] Berggvist D. Review of fondaparinux sodium injection for the prevention of venous thromboembolism in patients undergoing surgery. Vasc Health Risk Manag 2006;2:365—70.

CHAPTER 4

Antiplatelets and Anticoagulants

Contents

4.1 INTRODUCTION

Patients with ongoing or recent anticoagulant or antiplatelet therapy need to have coagulation status safely optimized for surgery or any other invasive procedure. Patients also need to be reverted to anticoagulation after intervention. To achieve an acceptable balance between coagulation and the risk of thrombosis, the attending physician needs to have an in–depth understanding of the data available in the literature regarding anticoagulants and antiplatelets.

This chapter discusses various antiplatelets and anticoagulants that are available. The principal focus is on monitoring efficacy of therapy and strategies for reversal effect in case of bleeding or in anticipation of surgery.

Several anticoagulants and antiplatelet agents are used for patients undergoing surgery, and sometimes the effects of such medications are reversed by using reversal agents [1]. A summary of anticoagulants and antiplatelets, together with their effects, administration routes, times to discontinue prior to surgery and available counteragents, is provided in Table 4.1. In the following sections, more details of these individual agents are provided.

A. Nguyen, A. Dasgupta, A. Wahed:
Management of Hemostasis and Coagulopathies for Surgical and Critically Ill Patients.
DOI: http://dx.doi.org/10.1016/B978-0-12-803531-3.00004-5

Table 4.1 Common anticoagulants and antiplatelets

Medication	Effect	Administration	Time to be discontinued prior to surgery	Counteragents if not discontinued adequately prior to surgery
Coumadin	Vitamin K antagonist	Oral	4–7 days	Vitamin K: 1–10 mg (oral 24 h or more before surgery; IV 6–24 h; FFP <6 h); PCC for patients with poor volume tolerance
Unfractionated heparin	AT inhibition	IV	4–6 h	Protamine: 1 mg per 100 U (1 mg)
LMWH	Indirect FXa inhibition	IV	24 h	Protamine: 1 mg per 100 anti-Xa units of LMWH (only 66% neutralized)
Clopidogrel	ADP-P2Y12 inhibition	Oral	5 days	None (platelets during or after surgery)
Ticagrelor (Brilinta)	ADP-P2Y12 inhibition	Oral	<5 days	None (platelets during or after surgery)
Prasugrel	ADP-P2Y12 inhibition	Oral	7 days [2]	None (platelets during or after surgery)
Abciximab (ReoPro)	Glycoprotein IIb/IIIa inhibition	IV	24–48 h [3,4]	None (platelets during or after surgery)
Eptifibatide (Integrilin)	Glycoprotein IIb/IIIa inhibition	IV	4–6 h [4]	None (platelets during or after surgery)
ASA	Cyclooxygenase inhibition	Oral	5–7 days (no discontinuation prior to cardiovascular surgery)	None (platelets during or after surgery)

Drug	Mechanism	Route	Timing	Reversal
Dabigatran (Pradaxa)	Direct thrombin inhibition	Oral	Normal or mild impairment (CrCl >50 mL/min): 3 days Moderate impairment (CrCl 30–50 mL/min): 4–5 days	Dialysis, FVIIa (not well validated) [5], PCC
Bivalirudin (Angiomax)	Direct thrombin inhibition	IV	2 hr	Dialysis, PCC, FVIIa (not well validated) [6]
Rivaroxaban	Direct FXa inhibition	Oral	Normal or moderate impairment (CrCl >30 mL/min): 3 days Severe impairment (CrCl 15–29.9 mL/min): 4 days	PCC
Apixaban	Direct FXa inhibition	Oral	Normal or mild impairment (CrCl >50 mL/min): 3 days Moderate impairment (CrCl 30–50 mL/min): 4 days	PCC

ASA, aspirin; CrCl, creatinine clearance; FFP, fresh frozen plasma; IV, intravenously; LMWH, low-molecular-weight heparin; PCC, prothrombin complex concentrate.

Table 4.2 INR therapeutic range for Coumadin

INR	Therapeutic intensity range
<1.5	—
1.5−2.0	Low
2.0−3.0	Moderate
2.5−3.5	High
3.0−4.0	Very high
5.0	Critical value
10.0	Upper reportable limit

4.2 MONITORING AND REVERSAL FOR COUMADIN

Coumadin is a brand name for warfarin, an anticoagulant used for prevention of thrombosis and thromboembolism. This drug inhibits vitamin K epoxide reductase, the enzyme that recycles oxidized vitamin K_1 to its reduced form. Warfarin is a synthetic drug that was approved for use in 1954; it is a racemic mixture of two optical isomers, with S-warfarin having five times the potency of R-isomer. The therapeutic range for international normalized ratio (INR) with patients on Coumadin is listed in Table 4.2.

The required Coumadin dose is variable and dependent on a number of patient-specific and environmental factors [7]. A patient's need for Coumadin should be carefully evaluated in order to determine Coumadin sensitivity. Guidelines for initial dosing of Coumadin are provided in Table 4.3, and guidelines for maintenance dosage of Coumadin are provided in Table 4.4. Important considerations for Coumadin therapy include the following:

- Baseline INR is recommended prior to initiating Coumadin therapy to assess sensitivity.
- An INR within the past 48 h is acceptable as a current baseline INR.
- Patients should be carefully monitored with each dose, and adjustments in dose are required based on INR values.
- With initial dosing, the INR will usually increase within 24−36 h.
- Daily INR should be obtained in hospitalized patients being initiated on Coumadin until INR is within the desired therapeutic range, after which INR can be evaluated twice weekly.

4.2.1 Reversal of Coumadin With Supratherapeutic INR

Sometimes patients on Coumadin show elevated INR, and it is important to reverse the effect of Coumadin in such patients to minimize risk of

Table 4.3 Initial Coumadin dosing guidelines

Day	INR	Coumadin high sensitivity (mg)[a]	Coumadin moderate sensitivity (mg)[b]	Coumadin low sensitivity (mg)[c]
Day 1	Baseline INR	2—5	5	7.5
Day 2	<1.5	2—5	5	5—7.5
	1.5—1.9	2	2.5	2.5—5
	2—2.5	1—2	1—2.5	1—2.5
	>2.5	None	None	None
Day 3	<1.5	5—10		
	1.5—1.9	2.5—5		
	2—2.5	0—2.5		
	2.6—3	0—2.5		
	>3	None		
Day 4	<1.5	10		
	1.5—1.9	5—7.5		
	2—3	2.5—5		
	>3	0—2.5		
Day 5	<1.5	10		
	1.5—1.9	5—7.5		
	2—3	2.5—5		
	>3	0—2.5		
Day 6	<1.5	7.5—12.5		
	1.5—1.9	5—10		
	2—3	2.5—5		
	>3	0—2.5		
Day 7	Make adjustment based on total weekly dose (increase or decrease) by 5—20%, depending on current INR and target INR			

[a]High Coumadin sensitivity: Baseline INR >1.5, age >65 years, significant hepatic disease, decompensated congestive heart failure, malnourished, malabsorption syndrome/chronic diarrhea, cancer, hypoalbuminemia (<2), thyrotoxicosis, genetic polymorphism of cytochrome P450 2C9.
[b]Moderate Coumadin sensitivity: Baseline INR 1.2—1.5, age 50—65 years, concurrent CYP450 enzyme inhibitor.
[c]Low Coumadin sensitivity: Baseline INR <1.2, age <50 years, no other risk factors.

bleeding. Guidelines for management of patients with supratherapeutic INR are listed in Table 4.5.

If the reversal of Coumadin effect prior to surgery is needed, the following guidelines may be used:
- Goal is for INR <1.5
- For 24 h or more before surgery: 2.5—5.0 mg oral vitamin K

Table 4.4 Maintenance Coumadin dosing guidelines

For target INR 2.0−3.0
INR < 2.0: Increase weekly dose by 10−15%
INR 2.0−3.0: No change
INR 3.1−3.5: Decrease weekly dose by 5−15%
INR 3.6−4.0: Hold 0−1 dose; then decrease weekly dose by 10−15%
INR >4.0: Hold dose until INR therapeutic; assess bleeding risk ± vitamin K administration; decrease weekly dose by 10−20%

For target INR 2.5−3.5
INR < 2.5: Increase weekly dose by 10−15%
INR 2.5−3.5: No change
INR 3.6−4.0: Decrease weekly dose by 5−15%
INR 4.1−5.0: Hold 0−1 dose; then decrease weekly dose by 10−15%
INR >5.0: Hold dose until INR therapeutic; assess bleeding risk ± vitamin K administration; decrease weekly dose by 10−20%

- For 6−24 h: 2.5−5.0 mg vitamin K intravenously (IV)
- For less than 6 h: fresh frozen plasma (FFP) 10 mL/kg (typically two FFPs)
  However, for patients with a high risk of thrombosis requiring Coumadin before surgery [ie, patients with a left ventricular assist device (LVAD), total artificial heart, mechanical heart valve, prior stroke, intracardiac thrombus, cardioembolic events, etc.], it is important that the INR before the procedure be higher but not supratherapeutic. INR can be kept at approximately 2.0 by the time of the procedure.

Prothrombin complex concentrate (PCC; brand name: KCentra) is reserved for select patients, such as those with poor tolerance for volume overload. KCentra is prepared from fresh frozen human plasma and contains clotting factors II, VII, IX and X, as well as protein C and S. KCentra may be used for reversal of Coumadin anticoagulation immediately prior to emergent surgeries, including cardiovascular surgeries. Dosing guidelines are as follows:

- Dosing (use maximum weight of 100 kg if patient >100 kg)
- INR between 1.5 and 3.99: Recommended dosage is 25 units/kg
- INR between 4 and 5.99: Recommended dosage is 35 units/kg
- INR ≥ 6: Recommended dosage is 50 units/kg
- No more KCentra should be used after this dose (50 units/kg). Repeat dosing is not supported by clinical data and is therefore not recommended. Thus, the maximum cumulative dose is 50 units/kg per day

Table 4.5 Management of supratherapeutic INR with patients on Coumadin

INR	Symptom	Recommendations
Target range $<$ INR $<$ 5	No significant bleeding	Lower or omit dose, resume therapy at a lower dose when INR therapeutic. Monitor INR more frequently
5 $<$ INR $<$ 9	No significant bleeding	Omit next 1 or 2 doses and monitor INR. Resume at a lower dose when INR in target range. Consider vitamin K 1–2.5 mg orally, particularly if at increased risk of bleeding
9 $<$ INR	No significant bleeding	Hold Coumadin. Consider vitamin K 1, 2.5, 5 or 10 mg orally, particularly if at increased risk of bleeding. With vitamin K 5–10 mg, expect INR to be reduced substantially by 24–48 h. Monitor more frequently. Resume therapy at lower dose when INR therapeutic

Any INR elevation	Significant bleeding	• Hold Coumadin
		• Give vitamin K by slow IV infusion (10 mg IV/NS 50 mL over 30 min)
		• Supplemented with FFP or PCC:
		  − For INR 2–3.99: 10 mL/kg (type 2 FFP); also consider PCC, 25 units/kg
		  − For INR 4–5.99: 15 mL/kg (type 4 FFP); PCC is preferred, 35 units/kg
		  − For INR $\geq$ 6: 20 mL/kg (type 6 FFP); PCC is preferred, 50 units/kg
		• PCC required for patients with poor tolerance to volume loading
		• Recombinant factor VIIa may be considered as a last resort (10–15 μm/kg) rounded to the nearest milligram, intravenous administration over 2 min

It is important to consider that vitamin K (5—10 mg) also needs to be administered to prevent Coumadin rebound after the effect of KCentra has weaned off in the post-op period. Another dose of vitamin K (5—10 mg) may also be considered 10—12 h later. Other important issues include the following:

- For heart transplant surgery, PCC dose to be mixed and administered only after visualization confirmed by harvesting surgeons. Reversal too soon would require patients to be on heparin and hence would take longer to achieve therapeutic INR with Coumadin if administered later.
- Another consideration is that KCentra contains heparin. Therefore, it is contraindicated in patients with heparin-induced thrombocytopenia (HIT). It can only be infused during surgery with appropriate measures (typically plasma pheresis prior to surgery and heparin neutralization after cardiopulmonary bypass).

There are other contraindications for use of KCentra if the patient has one of the following conditions:

- Suspected disseminated intravascular coagulation
- Acute myocardial infarction, acute septicemia, acute crush injury, acute peripheral arterial occlusion, acute thrombotic stroke, acute deep vein thrombosis (DVT) or pulmonary embolism (within 3 months) or high-risk thrombophilia
- Lupus anticoagulant/anticardiolipin antibodies
- Protein C, protein S or antithrombin deficiency
- Homozygous factor V Leiden
- Double heterozygous (factor V Leiden/factor II G20210A prothrombin mutation)
- Pregnancy
- In the past 30 days: history of transient ischemic attack, angina pectoris or limb claudication

4.3 MONITORING AND REVERSAL FOR UNFRACTIONATED HEPARIN

The therapeutic range for unfractionated heparin (UFH), anti-Xa with patients on UFH is typically 0.3—0.7 IU/mL. A typical infusion protocol for acute coronary syndrome is shown in Table 4.6.

Table 4.6 Typical infusion protocol for heparin weight-based acute coronary syndrome (therapeutic PTT of 60−80 s, corresponding to UFH anti-Xa of 0.3−0.7 IU/mL)

PTT range (s)	Action
<30	Increase rate by 3 units/kg per hour (heparin dosing weight)
	Notify physician stat
	Draw PTT in 6 h
30−45	Increase rate by 2 units/kg per hour (heparin dosing weight)
	Draw PTT in 6 h
46−59	Increase rate by 1 unit/kg per hour (heparin dosing weight)
	Draw PTT in 6 h
60−80	Therapeutic
	Redraw PTT in 6 h to confirm
	PTT daily if three consecutive therapeutic PTT values
81−90	Decrease rate by 1 unit/kg per hour (heparin dosing weight)
	Draw PTT in 6 h
91−99	Hold infusion for 30 min
	Decrease rate by 2 units/kg per hour (heparin dosing weight)
	Notify physician stat
	Draw PTT in 6 h
>99	Hold infusion for 60 min
	Decrease rate by 3 units/kg per hour (heparin dosing weight)
	Notify physician stat
	Draw PTT in 6 h

Critical points in monitoring therapy with UFH include the following:

- Starting dose typically at 4−6 units/kg per hour.
- Some patients may be at risk for both bleeding and thrombosis manifested with recent episodes of both; a modified anticoagulant protocol may be needed.
- In general, elevated levels of hemoglobin, bilirubin or lipids (results from hemolyzed, icteric or lipemic specimens) may interfere with the chromogenic assay for anti-Xa and yield a spurious decrease in the measured level [2]. The spurious levels result from free hemoglobin >1.5 g/L (150 mg/dL), direct bilirubin >342 μmol/L (20 mg/dL), indirect bilirubin >236 μmol/L (14 mg/dL) and triglycerides >8 mmol/L (708 mg/dL). Often, the result will not be available when the interference substances exceed the thresholds, and an error message is given.

- Partial thromboplastin time (PTT) on the coagulation instruments with clot-based tests (mechanically measuring the clot) is not affected by elevated levels of hemoglobin, bilirubin or lipids.
- For patients without lupus anticoagulant and without factor deficiencies, anti-Xa−UFH is also appropriate to monitor UFH. Anti-Xa−UFH and PTT are well correlated.
- For patients with lupus anticoagulant, PTT is not suitable to monitor UFH because the patients' baseline PTT is prolonged with no risk of bleeding. The standard of care is the use of anti-Xa−UFH.
- For patients with clotting factor deficiency (including liver disease patients and vitamin K deficiency), this presents a problem because increasing UFH dose to a therapeutic anti-Xa−UFH level can cause bleeding [3]. The reason is that a therapeutic level of anti-Xa−UFH would be supratherapeutic for patients with underlying coagulation factor deficiency. Anti-Xa−UFH level measures only the UFH level in the plasma. PTT, on the other hand, is a global test that takes into account the effects of UFH and any factor deficiency. Patients with liver problems and other coagulopathies are at risk for bleeding when using anti-Xa−UFH to dose UFH. PTT should be in the supratherapeutic range in such cases.
- Most anti-Xa−UFH reagents do not contain antithrombin (AT) and use the patient's AT for the reaction; a low AT level (<60%) would underestimate the heparin concentration. However, the anti-Xa−UFH level in this case reflects the effective level of heparin (comparable to PTT, which is subtherapeutic with risk for clotting). In this case, an attempt must be made to raise it (either with more UFH infusion or by giving FFP/AT concentrates if AT is too low).

4.3.1 Reversal of UFH

Generally, 1 mg of protamine is used to neutralize 1 mg (100 units) of UFH. For a man with a body weight of 70 kg (which requires 21,000 units of UFH with 300 units/kg dosage in cardiopulmonary bypass), the required amount is 210 mg protamine. For patients with bleeding not in the operating room but having prolonged PTT/thrombin time (TT) due to heparin, heparin correction can be started with 25 mg protamine sulfate IV, which should neutralize 2500 units of UFH. Then PTT/TT should be monitored to determine if further dose is needed. PTT/TT should be normalized readily, within 10−15 min, with quick neutralization of heparin by protamine sulfate. Protamine needs to be infused at least over 10 min. A faster rate may cause hypotension or ana-phylactoid reaction. If a patient's mean arterial pressure (MAP) is low, infusion should be done over 30 min with close monitoring of MAP.

4.4 MONITORING AND REVERSAL FOR BIVALIRUDIN (ANGIOMAX)

A typical bivalirudin anticoagulant protocol for HIT is shown in Table 4.7 with a therapeutic PTT of 60—80 s. Important points involving monitoring bivalirudin therapy as well as reversal of effects include the following:

- An initial dose for bivalirudin is typically 0.05 mg/kg per hour.
- Monitoring of PTT should be done every 4 h until the PTT value is within therapeutic range and also 4 h following any rate change.
- After two consecutive PTT values in therapeutic range, further monitoring of PTT can be conducted every 12 h while the patient is on bivalirudin.
- An infusion rate less than 0.01 mg/kg per hour does not prolong PTT to any noticeable degree. Only a rate of 0.01 mg/kg per hour or higher prolongs PTT.
- Bivalirudin also slightly prolongs prothrombin time (PT) (typically up to ~18 s).
- Some patients may be at risk for both bleeding and thrombosis manifested with recent episodes of both; a modified anticoagulant protocol may be needed.
- Dilute thrombin time (DTT) is useful to monitor bivalirudin in patients with lupus anticoagulant [4]. PTT is not suitable to monitor bivalirudin because these patients' baseline PTT is prolonged with no risk of bleeding. A typical therapeutic DTT range for bivalirudin is 60—100 s.

Table 4.7 Typical Bivalirudin anticoagulant protocol for HIT

PTT range (s)	Action
<60	Increase bivalirudin rate by 20%
60—80	None (therapeutic range)
81—90	Reduce bivalirudin rate by 25%
91—105	Reduce bivalirudin rate by 50%
>105	Hold bivalirudin for 1 h and reduce rate by 50%
>200	Hold bivalirudin for 2 h and then check stat PTT. Resume dosing per protocol with repeat PTT

- For patients without lupus anticoagulant and without factor deficiencies, DTT is also adequate to monitor bivalirudin. PTT and DTT are well correlated in this case.
- For patients with clotting factor deficiency (including liver disease patients or patients with vitamin K deficiency), it presents a problem because setting bivalirudin dose to a therapeutic DTT can cause bleeding. The reason is that a therapeutic level of DTT would be supratherapeutic for patients with underlying coagulation factor deficiency. DTT measures only the TT in diluted sample and does not show any deficiency preceding thrombin in the coagulation cascade. PTT, on the other hand, is a global test that takes into account the effect of bivalirudin and any factor deficiency in the common and intrinsic pathways. Patients with liver problems and other coagulopathies are at risk for bleeding using DTT to monitor and dose bivalirudin. PTT is in the supratherapeutic range in such cases.
- For highly suspected HIT (eg, significant drop in platelet count without other obvious etiologies, recent exposure to UFH such as in cardiopulmonary bypass, and either currently on UFH or not), the following is initiated: Bivalirudin anticoagulation for HIT with typical PTT target range of 50–80 sec; monitor for signs of bleeding on this regimen; avoid all heparins, including heparin flushes, and fractionated heparin until HIT Ab enzyme-linked immunosorbent assay (ELISA) results are available.
- If HIT Ab-ELISA is positive or the patient has clinical signs of thrombosis, continue bivalirudin until platelet count reaches $150,000\,\mu L^{-1}$, at which point Coumadin may initiated. Coumadin therapy should be started with low maintenance doses (eg, 2.5–5 mg) rather than a higher initial dose. Bivalirudin should be continued after the platelet count has reached a stable plateau and INR has reached the intended target range, after a minimum overlap of at least 5 days.
- A very low platelet count ($<15,000/\mu L$) is associated with a high risk of bleeding. Platelet transfusion can be considered for patients not currently on heparin and without signs of thrombosis.
- For patients with a need for DVT prophylaxis and who have a drop in platelet count with associated concern for HIT, DVT prophylaxis can be started with bivalirudin at a fixed low rate (eg, 0.005 mg/kg per hour). Another option for DVT prophylaxis with suspected HIT is fondaparinux 2.5 mg/day subcutaneously pending HIT Ab-ELISA results. If HIT Ab-ELISA is negative, prophylaxis can be switched to

UFH sq. If HIT Ab-ELISA is positive or the patient has clinical signs of thrombosis, bivalirudin protocol for HIT is ordered, with a typical PTT target range of 50–80 s.

- Delayed-onset HIT may occur days or even a few weeks after UFH exposure. This syndrome is associated with high-titer HIT antibodies that activate platelets even in the absence of pharmacologic heparin [5].
- PTT returns to normal approximately 1 h after discontinuation of bivalirudin (in therapeutic dose). It is safe to discontinue the medication 2 h prior to surgery in patients with normal renal function. Bivalirudin significantly prolongs thromboelastography (TEG) R and slightly decreases TEG alpha. Dialysis and FVIIa have been used to reverse bivalirudin in bleeding patients. FVIIa should be started at low dose (15–30 µg/kg).

4.5 MONITORING AND REVERSAL FOR DABIGATRAN

DTT would be useful to monitor dabigatran. A typical normal range for DTT with dabigatran is 55–110 s. A normal TT indicates that dabigatran is no longer effective. A normal PTT indicates a low level of dabigatran (most of the drug has been cleared). Dabigatran does not have antidote, but dialysis does remove this agent (60% removal over 2 or 3 h) [6].

PCC can be used for bleeding patients on dabigatran. Dose is typically 35 units/kg and is restricted to a single dose. (Use maximum weight of 100 kg if patient >100 kg.) The maximum cumulative dose is 50 units/kg per day. Use is restricted to a single dose. Repeat doses do not improve efficacy and increase risk for thromboembolic complications. FVIIa at low dose (15–30 µg/kg) has been suggested, but it is not well validated.

4.6 MONITORING AND REVERSAL FOR RIVAROXABAN (XARELTO) AND APIXABAN

Monitoring as well as reversal of action of rivaroxaban are important clinically. Anti-Xa level (for low-molecular-weight heparin) can be ordered to approximately measure the activity of the direct oral Xa inhibitors rivaroxaban and apixaban. However, no thresholds for bleeding risk have been established. The following are important points to remember:

- PT/PTT are not useful to assess apixaban level even though a normal PT indicates that rivaroxaban is at a low level (most of the drug has been cleared). INR is not the appropriate dosing guide for anti-Xa inhibitor.

- If PT/PTT are more prolonged in patients taking rivaroxaban or apixaban, mixing study is needed. Depending on the results of the mixing study, further testing to rule out inhibitor (lupus anticoagulant) or factor deficiency (factor assays) is needed prior to surgery.
- Severe bleeding may be associated with the use of rivaroxaban and apixaban, requiring a reversal agent. Dialysis is not useful to clear rivaroxaban or apixaban. PCC may be useful for bleeding associated with these medications. Dose is typically 35 units/kg and is restricted to a single dose. (Use maximum weight of 100 kg if patient >100 kg.) The maximum cumulative dose is 50 units/kg per day. Use is restricted to a single dose. Repeat doses do not improve efficacy and increase risk for thromboembolic complications.

4.7 MONITORING FOR ASPIRIN AND P2Y12 PLATELET INHIBITORS

Aspirin along with P2Y12 platelet inhibitors (clopidogrel, prasugrel and ticagrelor) are clinically used to avoid clot formation. In Table 4.8, platelet function testing results for patients on aspirin (acetylsalicylate; ASA) and/or P2Y12 platelet inhibitors are summarized. Important considerations for monitoring aspirin and ADP PY12 inhibitors include the following:

- Platelet aggregation with ADP is affected by many conditions besides ASA and P2Y12 inhibitors (eg, platelet storage pool disease and uremia). The patient's results need to be interpreted with this caveat.
- Factors causing inaccurate results (or no reading) by VerifyNow include the following:
  - Severe anemia, severe thrombocytopenia, Greiner tube filled below or above the halfway level, no discard sample prior to Greiner sample, a purple-top tube filled before Greiner tube and Greiner tube not hand-carried to lab. If a test result is unexpectedly low, repeat testing, ensuring that all technical artifacts are prevented.
  - Patients who have been treated with glycoprotein IIb/IIIa inhibitor drugs should not be tested for 14 days after discontinuation of drug administration for abciximab (ReoPro) and up to 48 h for eptifibatide (Integrilin) and tirofiban (Aggrastat). The recovery time varies among individuals and is longer for patients with renal dysfunction. VerifyNow P2Y12 testing with residual effect of

Table 4.8 Platelet function testing results for patients on ASA and/or ADP P2Y12 inhibitors: Typical results for therapeutic ranges

Medication	Platelet aggregation		VFN	
ASA only	AA (5 mg/mL)	<15%	VFN-ASA	<550 ARU
	ADP (2.5 μM/mL)	20–50%	VFN-P2Y12	>210 PRU
	ADP (50 μM/mL)	>60%		
ADP P2Y12 inhibitor only	AA (5 mg/mL)	>60%	VFN-ASA	>550 ARU
	ADP (2.5 μM/mL)	<40%	VFN-P2Y12	<210 PRU
	ADP (50 μM/mL)	<50%		
ASA and ADP P2Y12 inhibitor	AA (5 mg/mL)	<15%	VFN-ASA	<550 ARU
	ADP (2.5 μM/mL)	<40%	VFN-P2Y12	<210 PRU
	ADP (50 μM/mL)	<50%		

AA, arachidonic acid; ADP, adenosine diphosphate; ARU, aspirin response units; PRU, Plavix reactive units; VFN, VerifyNow; VFN-ASA, VerifyNow for Aspirin; VFN-P2Y12, VerifyNow for ADP P2Y12 inhibitors.

these medications will give an error signal or will give a falsely low level. Use platelet function screen with ADP to evaluate platelet function instead (see the following two entries).

- Platelet function screen with ADP (abbreviated platelet aggregation study with only two ADP concentrations of 2.5 and 50 μM/mL) can be ordered stat with turnaround time of approximately 1 h during the day shift (07:00 am to 03:00 pm) at our institution.
- Platelet aggregation in patients who have been treated with glycoprotein IIb/IIIa inhibitor drugs typically shows flat lines similar to those of Glanzmann thrombasthenia. Platelet function screen with ADP (rather than VFN-P2Y12) should be used to evaluate platelet function. Response to 50 μM/mL ADP typically returns to normal within 48 h for ReoPro [8] and 6 h for Integrilin. The recovery time is longer for patients with renal dysfunction.
- Clinically important thresholds for VerifyNow P2Y12 test:
  - High risk of bleeding with surgery: <208 Plavix reactive units (PRU) [9]
  - Acceptable risk of bleeding with surgery: >208 PRU [9]
  - High risk of thrombosis after stent placement: >230 PRU (resistance to P2Y12 inhibitors)
  - Low risk of thrombosis after stent placement: <208 PRU (therapeutic range)

4.8 BRIDGING ANTICOAGULANTS FOR SURGERY

The following are major points dealing with the use of anticoagulants in surgery:

- Coumadin should be discontinued at least 5 days before surgery.
- Monitoring INR closely pre- and postprocedure to time bridging therapy is necessary.
- Therapy with UFH should be initiated when INR falls below 2.0. Then UFH dose may be increased to achieve PTT value within therapeutic range.
- If the INR remains elevated (>2.0) 1 or 2 days before surgery, administration of vitamin K 1−2 mg orally is highly recommended. If the patient is scheduled to undergo surgery in less than 6 hr, dosage should be vitamin K 2.5 mg IV.
- Discontinuation of UFH 4 h before surgery is essential.
- For low-bleeding-risk surgery, therapy with UFH can be resumed after 24 h provided hemostasis has been achieved.
- For major- or high-bleeding-risk patients, after surgery, resumption of therapeutic dose of UFH should be delayed for 48−72 h, or the administration of low-dose UFH should be considered based on bleeding risk and adequacy of postoperative hemostasis for each patient individually.
- Initiation of Coumadin therapy at the appropriate time is recommended, and such therapy can be continued until INR is at least 2.0. Then administration of UFH can be discontinued.
- Patients with a high risk of thrombosis (eg, patients with LVAD, total artificial heart, mechanical heart valve, prior stroke, intracardiac thrombus or cardioembolic events) may present difficult anticoagulant management problems due to risks of thrombosis versus risk of bleeding before and during surgery. Patients may need to be on Coumadin until the time of surgery. The INR value at which the risk of bleeding increases is unknown; it is assumed to be elevated when the INR is no more than 2.0 [10]. For patients with a high risk of thrombosis requiring Coumadin before surgery, it is important that the INR before the procedure not be supratherapeutic. INR may be kept at approximately 2.0 by the time of the procedure.

4.9 CONCLUSIONS

This chapter summarized the published literature concerning the pharmacokinetics of anticoagulants and antiplatelet medications that are currently available for clinical use and other aspects related to their management. The effects of each anticoagulant and antiplatelet medication, as well as the monitoring of anticoagulation/antiplatelet intensity, were described in detail. Moreover, the chapter described and discussed the clinical applications and optimal dosages of therapies, practical issues related to their initiation and monitoring, adverse events such as bleeding and other potential side effects and also available strategies for reversal.

REFERENCES

[1] Bracey AW, Reyes MA, Chen AJ, Bayat M, et al. How do we manage patients treated with antithrombotic therapy in the perioperative interval. Transfusion 2011;51:2066−77.

[2] STA liquid Anti-Xa reagent package inset. REF 00322US; November 2013.

[3] Price E, Jean J, Nguyen MH, Krishnan G, et al. Discordant aPTT and anti-Xa values and outcomes in hospitalized patients treated with intravenous unfractionated heparin. Ann Pharmacother 2013;47:151−8.

[4] Love JE, Ferrell C, Chandler WL. Monitoring direct thrombin inhibitors with plasma diluted thrombin time. Thromb Haemost 2007;98:234−42.

[5] Warkentin TE, Greinacher A. Heparin-induced thrombocytopenia and cardiac surgery. Ann Thorac Surg 2003;76:2121−31.

[6] Van Ryn J, Stangier J, Haertter S, Kiesenfeld KH, et al. Dabigatran etexilate—a novel, reversible, oral direct thrombin inhibitor: interpretation of coagulation assays and reversal of anticoagulant activity. Thromb Haemost 2010;103:1116−27.

[7] Ageno W, Gallus AS, Wittkowsky A, Crowther M, et al. Oral anticoagulant therapy: antithrombotic therapy and prevention of thrombosis, 9th ed: American College of Chest Physicians evidence-based clinical practice guidelines. Chest 2012;141(2 Suppl.): e44S−88S.

[8] REOPRO (abciximab) injection, solution. Daily Med. http://dailymed.nlm.nih. gov/dailymed/lookup.cfm?setid = 033d4c3b-4630-4256-b8f7-9ed5f15de9a3.

[9] Reed GW, Kumar A, Guo J, Aranki S, et al. Point of care platelet function testing predicts bleeding in patients exposed to clopidogrel undergoing coronary artery bypass grafting: verify pre-op TIMI 45-a pilot study. Clin Cardiol 2015;38:92−8.

[10] Douketis JD, Spyropoulos AC, Spencer FA, Mayr M, et al. Perioperative management of antithrombotic therapy: antithrombotic therapy and prevention of thrombosis, 9th ed: American College of Chest Physicians evidence-based clinical practice guidelines. Chest 2012;141(2 Suppl.):e326S−50S.

CHAPTER 5

Preoperative Assessment of Patients

Contents

5.1 INTRODUCTION

The preoperative assessment of hemostasis of a patient scheduled for surgery is important for the management of bleeding of the patient during and after surgery as well as to avoid thromboembolic events during this period. The tools that are routinely used for such an assessment usually include the following:

- History: This includes taking medical history with emphasis on clinical bleeding history, surgical history, list of medications currently taken by a patient and any prior transfusion history.
- Physical examination.

A. Nguyen, A. Dasgupta, A. Wahed:
Management of Hemostasis and Coagulopathies for Surgical and Critically Ill Patients.
DOI: http://dx.doi.org/10.1016/B978-0-12-803531-3.00005-7
91

- Laboratory testing: Laboratory tests ordered for patients include liver function panel, renal function panel, complete blood count, appropriate coagulation tests and blood banking tests.

5.2 HISTORY

Taking a thorough history of the patient is essential during preoperative assessment.

5.2.1 Medical History

It is important to enquire about history of hepatic or renal dysfunction because both are associated with increased risk of bleeding. Hepatic dysfunction may be responsible for decreased production of coagulation factors, and renal dysfunction is associated with platelet dysfunction. History of bleeding disorders or a family history of bleeding disorders should be enquired; even if no such disorders are found, this should be documented in the patient history. Questions regarding bleeding (eg, history of blood in stool or urine, bleeding into joints, excessive bleeding after trivial injury or tooth extraction, and menorrhagia) should be asked. An example of a questionnaire with directed questions is provided in Table 5.1.

5.2.2 Surgical History

If the patient has had previous surgeries, enquiries regarding excessive use of blood products should be made. This should point toward underlying problems such as bleeding disorders. Redo surgeries (eg, resternotomy

Table 5.1 Questionnaire

Q1	Do you experience frequent nosebleeds or excessive bleeding when brushing your teeth?
Q2	Have you seen blood in your stool or urine?
Q3	Have you had excessive bleeding with tooth extractions?
Q4	Have you had excessive menstrual blood loss?
Q5	Do you bruise easily? Do you have excessive bleeding with minor cuts or injuries?
Q6	Have you ever bled into muscles or joints?
Q7	Have you had excessive bleeding from procedures or surgeries?
Q8	Are you on blood thinners (such as antiplatelets or anticoagulants)? Are you taking OTC medicines or any alternative medicine?
Q9	Are you aware of any family member with bleeding disorders?

that presents with scar tissue complicating cardiac surgeries) are also associated with increased risk of bleeding.

5.2.3 Medication History

Therapy with antiplatelets and/or anticoagulants is an important cause for increased risk of bleeding during surgery. Detailed history of such medications should be documented. These products will need to be discontinued prior to surgery, and reversal if possible may be required for emergent surgeries. Various antiplatelets and anticoagulants that need to be discontinued prior to surgery are listed in Table 5.2.

5.2.4 Transfusion History

Enquiries should be made as to whether the patient received blood products in the past and, if so, if there were any untoward reactions. If there are any special requirements for blood products, these should be assessed at this time. Examples of these include requirement for cytomegaly virus-negative blood and irradiated cellular products.

5.3 PHYSICAL EXAMINATION

In the physical examination, the clinician should observe for the following:
- Petechiae and ecchymosis: May be seen in individuals with thrombocytopenia or thrombocytopathia
- Hematomas: May indicate clotting factor deficiencies
- Joint swelling or deformity: May indicate bleeding into joints
- Skin elasticity, hyperextensible joints: May be seen in collagen vascular diseases such as Ehlers–Danlos syndrome
- Jaundice: May indicate abnormal liver function
- Splenomegaly: May indicate chronic liver disease

5.4 LABORATORY TESTING

Several studies have demonstrated that for individuals with a negative history and physical examination, routine laboratory testing is unnecessary [1–4]. In other circumstances, laboratory testing should include liver function tests, creatinine, blood urea nitrogen (BUN) and complete blood count (CBC). Appropriate coagulation tests, such as prothrombin time (PT), partial thromboplastin time (PTT), platelet aggregation study,

Table 5.2 Antiplatelets and anticoagulants that need to be discontinued prior to surgery

Medication	Mechanism of action/effect	Route of administration	Recommended duration of discontinuation prior to surgery	Agents to counter actions, if required
Warfarin	Vitamin K antagonist	Oral	4–7 days	Vitamin K 1–10 mg (oral 24 h or more before surgery; IV 6–24 h; FFP less than 6 h)
Unfractionated heparin	AT inhibition	IV	4–6 h	Protamine: 1 mg per 100 units (1 mg)
LMWH	Indirect FXa inhibition	IV	24 h	Protamine: 1 mg per 100 anti-Xa units of LMWH (only 66% neutralized)
Clopidogrel	ADP-P2Y12 inhibition	Oral	5 days	None (platelets during or after surgery)
Ticagrelor (Brilinta)	ADP-P2Y12 inhibition	Oral	<5 days	None (platelets during or after surgery)
Prasugrel	ADP-P2Y12 inhibition	Oral	7 days	None (platelets during or after surgery)
Abciximab (ReoPro)	Glycoprotein IIb/IIIa inhibition	IV	24–48 h	None (platelets during or after surgery)
Eptifibatide (Integrilin)	Glycoprotein IIb/IIIa inhibition	IV	4–6 h	None (platelets during or after surgery)
ASA	Cyclooxygenase inhibition	Oral	No discontinuation	None (platelets during or after surgery)

Drug	Mechanism	Route	Discontinuation	Reversal
Dabigatran (Pradaxa)	Direct thrombin inhibition	Oral	Normal or mild impairment (CrCl >50 mL/min): 3 days Moderate impairment (CrCl 30–50 mL/min): 4 or 5 days	Dialysis, FVIIa (not well validated), PCC
Bivalirudin (Angiomax)	Direct thrombin inhibition	IV	2 h	
Rivaroxaban	Direct FXa inhibition	Oral	Normal or moderate impairment (CrCl >30 mL/min): 3 days Severe impairment (CrCl 15–29.9 mL/min): 4 days	Dialysis, PCC, FVIIa (not well validated) PCC
Apixaban	Direct FXa inhibition	Oral	Normal or mild impairment (CrCl >50 mL/min): 3 days Moderate impairment (CrCl 30–50 mL/min): 4 days	PCC

ASA, aspirin; CrCl, creatinine clearance; FFP, fresh frozen plasma; IV, intravenously; LMWH, low-molecular-weight heparin; PCC, prothrombin complex concentrate.

thromboelastograph (TEG) and von Willebrand panel, should be ordered depending on clinical evaluation of the patient. Appropriate uses of these tests are discussed in detail in the chapter "Coagulation-Based Tests and Their Interpretation."

5.4.1 Blood Bank Testing

Blood bank testing should start with a type and screen. Note that a type and screen is valid for 3 days. A negative screen denotes that the patient does not have preformed antibodies. A positive screen means that the blood bank will require additional time to determine the nature of the antibody; if the antibody is considered to be clinically significant, the blood bank will have to find corresponding antigen-negative blood. See the chapter "Blood Bank Testing and Blood Products" for an in-depth discussion on blood bank testing.

5.5 ANTIPLATELETS AND ANTICOAGULANTS: IMPACT ON SURGERY

Patients who require surgery are at times receiving antiplatelet and/or anticoagulants. Interruption of anticoagulants for surgery may increase the risk of thromboembolism. Continuing use of anticoagulants, on the other hand, increases the risk of bleeding. Both of these situations adversely affect mortality [5—8].

Agents such as warfarin, if stopped, require several days before the anticoagulant effect is reduced. However, the newer oral anticoagulants (direct thrombin inhibitors and Xa inhibitors) have shorter half-lives. These may be easier to discontinue. Unfortunately, these agents lack specific antidotes, and if patients who are on these drugs bleed or require surgery urgently, management may be difficult. The following approaches are reasonable to take when faced with patients on anticoagulants who require surgery:

- Assessment of thromboembolic risk: Individuals who are at high risk for thromboembolism should have the shortest possible anticoagulant-free period. Such patients are those with atrial fibrillation, prosthetic heart valves and recent (within the past 3 months) arterial or venous thromboembolism.
- Assessment of bleeding risk: Bleeding risk is determined by type and urgency of surgery. Surgeries such as coronary artery bypass surgery, renal biopsy and procedures lasting more than 45 min are considered

high-risk procedures for bleeding. In contrast, dental extractions and skin surgeries are associated with low risk of bleeding. High-risk surgeries definitely require interruption of anticoagulation. Patients with low risk can continue anticoagulation most of the time.

- Duration of discontinuation: This depends on the agent as well as the renal function if the agent is cleared by the kidneys.
- Bridging with another anticoagulant: This is typically done when a patient is on warfarin and is considered a high risk for thromboembolism. Warfarin is stopped, and heparin or low-molecular-weight heparin (LMWH) is administered while the effect of warfarin disappears. Heparin/LMWH can be stopped closer to the surgery date.

5.6 COMMONLY USED ANTIPLATELETS AND ANTICOAGULANTS

Commonly used anticoagulants are warfarin, heparin and LMWH, whereas commonly used antiplatelet agents are aspirin, P2Y12 inhibitors, glycoprotein IIb/IIIa inhibitors, thrombin inhibitors and factor Xa inhibitors. In this section, proper discontinuation of these agents, if required, prior to surgery is discussed.

5.6.1 Warfarin

Ideally, warfarin should be stopped 4—7 days prior to surgery. The effect of warfarin is measured by the international normalized ratio (INR). The INR prior to surgery should be <1.5. If surgery is required sooner, vitamin K, fresh frozen plasma (FFP) and prothrombin complex concentrate (PCC) may be used to reverse the effect of warfarin. The recommendations for the previously mentioned agents are summarized in Table 5.3.

Table 5.3 Reversal of warfarin prior to emergent surgery

Time before surgery	Reversing agent
>24 h	2.5—5.0 mg oral vitamin K
6—24 h	2.5—5.0 mg vitamin K IV
<6 h	FFP (one dose = 2 units; 10 mL/kg)
<6 h with concern for volume overload	PCC

5.6.2 Heparin: Unfractionated and Low-Molecular-Weight Heparin

Heparin should be stopped 4—6 h prior to surgery, and LMWH should be stopped 24 h prior to surgery. The effect of heparin may be tested for PTT and thrombin time (TT). Heparin prolongs PTT and TT. If the fibrinogen level is normal, prolonged TT is most likely due to heparin. With the use of heparin, TEG R is prolonged and shortens by at least 50% with TEG heparinase. Note that heparin may produce artifacts in TEG values, such as low angle alpha and maximum amplitude (MA). The effect of LMWH is tested by LMWH anti-Xa activity. Both heparin and LMWH effects may be reversed by protamine.

5.6.3 Antiplatelet Drugs

The antiplatelets that are prescribed include aspirin, P2Y12 inhibitors and glycoprotein IIB/IIIa inhibitors. The gold standard for platelet function testing is platelet aggregation studies. VerifyNow is available for testing for aspirin and P2Y12 inhibitors.

Note that TEG is not affected by aspirin and P2Y12 inhibitors. Thus, patients on these medications will have a normal MA value despite demonstrating platelet dysfunction by other testing methods. Therapeutic ranges for platelet function testing results for patients on aspirin and/or adenosine diphosphate (ADP) P2Y12 inhibitors are listed in Table 5.4.

Patients who have been treated with glycoprotein IIb/IIIa inhibitor drugs should not be tested for 14 days after discontinuation of drug

Table 5.4 Platelet function testing results for patients on aspirin and/or ADP P2Y12 inhibitors: Typical results for therapeutic ranges

Medication	Platelet aggregation	VFN
Aspirin (ASA) only	AA (5 mg/mL) <15% ADP (2.5 µM/mL) 20—50% ADP (50 µM/mL) >60%	VFN-ASA <550 ARU VFN-P >210 PRU
P2Y12 inhibitor only	AA (5 mg/mL) >60% ADP (2.5 µM/mL) <40% ADP (50 µM/mL) <50%	VFN-ASA >550 ARU VFN-P <210 PRU
ASA and P2Y12 inhibitor	AA (5 mg/mL) <15% ADP (2.5 µM/mL) <40% ADP (50 µM/mL) <50%	VFN-ASA <550 ARU VFN-P <210 PRU

AA, arachidonic acid; ADP, adenosine diphosphate; ARU, aspirin response units; PRU, Plavix reactive units; VFN, VerifyNow; VFN-ASA, VerifyNow for Aspirin; VFN-P, VerifyNow Plavix assay.

administration for abciximab (ReoPro) and up to 48 h for eptifibatide (Integrilin) and tirofiban (Aggrastat). The recovery time varies among individuals and is longer for patients with renal dysfunction. VerifyNow P2Y12 testing with residual effect of these medications will give an error signal or will give falsely low levels. Platelet aggregation studies should be performed to assess the effects of these drugs. Platelet aggregation in patients who have been treated with glycoprotein IIb/IIIa inhibitor drugs typically shows flat lines similar to those of Glanzmann thrombasthenia. Alternatively, platelet function screen with ADP may be performed to evaluate platelet function. Here, platelet aggregation results are obtained with ADP only at low (eg, 2.5 µM/mL) and high (50 µM/mL) concentrations. Response to 50 µM/mL ADP typically returns to normal within 48 h for ReoPro and 6 h for Integrilin. The recovery time is longer for patients with renal dysfunction.

5.6.4 Bivalirudin (Angiomax) and Dabigatran

Bivalirudin and dabigatran are direct thrombin inhibitors. Both drugs prolong TT and PTT. Bivalirudin is given intravenously, and with normal renal function it can be stopped 2 h prior to surgery. Dabigatran is given orally, and in patients with normal renal function this drug should be stopped 3 days prior to surgery. Neither drug has an antidote, but dialysis has been used to remove these agents in the setting of bleeding patients.

Note that Angiomax significantly prolongs TEG R and slightly decreases TEG alpha.

5.6.5 Rivaroxaban (Xarelto) and Apixaban

Rivaroxaban and apixaban are direct Xa inhibitors and can be administered orally. There are no antidotes for these agents. Dialysis is not useful in clearing these agents. In individuals with normal renal function, these drugs need to be stopped 3 days prior to surgery. Anti-Xa level (for LMWH) can be ordered to approximately measure activity of rivaroxaban or apixaban. However, no thresholds for bleeding risk have been established.

5.7 CONCLUSIONS

The preoperative assessment of hemostasis of a patient scheduled for surgery is important to manage bleeding of the patient during and after surgery and at the same time minimize risk of thromboembolic events during this period. The tools that are routinely used for such an assessment include a detailed history, physical examination, and appropriate laboratory and blood bank testing.

Several studies have demonstrated that for individuals with a negative history and physical examination, routine laboratory testing is unnecessary. In other circumstances, laboratory testing should include liver function tests, creatinine, BUN and CBC. Appropriate coagulation tests, such as PT, PTT, platelet aggregation study, TEG and von Willebrand panel, should be ordered depending on clinical evaluation of the patient.

The decision regarding whether or not to discontinue antiplatelets and anticoagulants depends on the thromboembolic risk as well as the bleeding risk of the patient. Determination of the duration of discontinuation also depends on the particular agent as well as the renal function of the patient. Bridging with a second anticoagulant may be necessary in certain clinical situations.

REFERENCES

[1] Velanovich V. The value of routine preoperative laboratory testing in predicting postoperative complications: a multivariate analysis. Surgery 1991;109:236.
[2] Narr BJ, Warner ME, Schroder DR, Warner MA. Outcomes of patients with no laboratory assessment before anesthesia and surgical procedure. Mayo Clinic Proc 1997;72:505.
[3] Velanovich V. Preoperative laboratory screening based on age, gender and concomitant medical diseases. Surgery 1994;115:56.
[4] Wattsman TA, Davies RS. The utility of preoperative laboratory testing in general surgery patients for outpatient procedures. Am Surg 1997;63:81.
[5] Douketis JD. Perioperative management of patients who are receiving warfarin therapy: an evidence based and practical approach. Blood 2011;117:5044.
[6] Spyropoulos AC, Douketis JD. How I treat anticoagulated patients undergoing an elective procedure or surgery. Blood 2012;120:2954.
[7] Torn M, Rosendaal FR. Oral anticoagulation in surgical procedures risks and recommendations. Br J Haematol 2003;123:676.
[8] Jaffer AK. Perioperative management of warfarin and antiplatelet therapy. Cleve Clin J Med 2009;76(Suppl. 4):S37.

CHAPTER 6

Intraoperative and Postoperative Assessment and Management of Coagulopathy

Contents

6.1 INTRODUCTION

Several clinical laboratory tests are performed during the intraoperative period with a goal of minimizing bleeding during surgery. Proper interpretations of test results is necessary to achieve such goal. In general, coagulopathy during most surgeries can be avoided using proper preoperative assessment (see the chapter "Preoperative Assessment of Patients"). As a result, intraoperative assessment of coagulopathy using laboratory tests may not be required for a majority of patients. Nevertheless, despite preoperative assessment, some patients with high risk for coagulopathy during surgery may require intraoperative assessment of coagulopathy as part of standard patient management. These high–risk patients include those undergoing major cardiovascular surgery, such as left ventricular assist device (LVAD) placement and multiple valve repair. These patients are at a higher risk of bleeding during surgery due to multiple factors, including use of high unfractionated heparin (UFH) dose during cardiopulmonary bypass (CPB) as well as a high incidence of liver and renal

A. Nguyen, A. Dasgupta, A. Wahed:
Management of Hemostasis and Coagulopathies for Surgical and Critically Ill Patients.
DOI: http://dx.doi.org/10.1016/B978-0-12-803531-3.00006-9

disorders. As a result, intraoperative protocol for laboratory assessment and management of coagulopathy in patients undergoing LVAD surgery is helpful to manage bleeding in such patients. In our hospital, we have successfully implemented such protocol.

In this chapter, essential aspects of coagulopathy in CPB are discussed, followed by discussion of LVAD protocols at our institution, including a summary of preoperative assessment for LVAD surgery. Next, details on intraoperative management provided. Miscellaneous information on the intraoperative management of coagulopathy is also provided. Finally, postoperative assessment and management of coagulopathy, an important part of our service, is discussed.

6.2 INTRAOPERATIVE ASSESSMENT AND MANAGEMENT OF COAGULOPATHY

Cardiac surgery with CPB exposes patients to increased risk of excessive perioperative blood loss resulting in transfusion of blood products. The particular type of procedure and the duration of CPB have a significant impact on bleeding risk. Although bleeding with CPB may be due to preexisting hemostatic abnormalities prior to surgery, CPB by itself may cause significant hemostatic alterations that predispose patients to coagulopathy.

6.2.1 Major Mechanisms of Bleeding Associated With Cardiopulmonary Bypass

Major mechanisms of bleeding known to be associated with CPB include the following:

- Crystalloid solution, which is used to prime the CPB circuit and also used as a component of cardioplegia, can result in substantial hemodilution and accounts for the decreased level of coagulation factors and platelets.
- Excessive activation of the hemostatic system during CPB may lead to consumption of platelets and coagulation factors, especially labile coagulation factors. Excessive hemostatic activation is due to activation of the contact factors in the extrinsic pathway with the extensive CPB surface and also to activation of the intrinsic pathway secondary to surgical trauma or retransfusion of pericardial blood.
- Excessive fibrinolysis may be triggered via CPB-mediated activation of factor XII and thrombin. Excessive plasmin activity may lead to hypofibrinogenemia; platelet dysfunction secondary to fibrinogen/fibrin

degradation product; and degradation of factors V, VIII and XIII. Plasmin may also cause proteolytic removal of platelet membrane glycoprotein Ib, impairing response of platelets to various agonists.

- Heparin may inhibit coagulation and platelet function. Similarly, excess protamine may inhibit coagulation and affect platelet function.

Many hemostatic abnormalities have been reported in patients with CPB, such as decreases in plasma coagulation factors, disseminated intravascular coagulation (DIC), isolated primary fibrinolysis, thrombocytopenia and transient platelet dysfunction. The latter two are considered to be the most important abnormalities in the early postoperative period with CPB.

Blood component administration in patients with excessive microvascular bleeding after CPB has been generally empiric and not based on direct assessment of hemostasis because the turnaround time (TAT) of laboratory-based tests is typically too long to be useful for acute management. Thus, transfusion of erythrocytes, platelets and fresh frozen plasma (FFP) to surgical patients requiring CPB varies considerably among institutions. FFP and platelets are frequently administered in an attempt to distinguish between microvascular bleeding related to hemostatic system impairment and surgical bleeding. This approach appears to be inefficient and is therefore inadequate for management of coagulopathy.

Complications of bleeding during and after cardiac surgery are significant. The increased requirement for blood transfusions places patients at risk for alloimmunization, nosocomial infections, transfusion-related infectious diseases and other transfusion-associated complications [1−3]. Bleeding patients may require additional surgeries with resternotomy, which occur in as many as 60% of LVAD patients [4]. Blood product usage increases the risk of human leukocyte antigen alloimmunization, making more difficult the process of finding an appropriately matched organ donor and placing patients at risk for future allograft rejection [2,5]. Large volumes of transfused blood products may trigger an inflammatory response that causes respiratory problems, pulmonary hypertension and the risk of right ventricular failure [4]. Inappropriate management of the bleeding patient can also cause poor outcomes by dilution of coagulation factors. Finally, cardiac surgeries use as much as 25% of the blood products in the United States [6]. Interventions that safely reduce bleeding and the resulting blood product use have the potential to improve patient morbidity with tremendous potential for cost savings.

6.2.2 Coagulation-Based Hemotherapy Service

One of the significant challenges to optimize blood product administration in order to manage excessive bleeding associated with cardiac surgery is the long TAT of laboratory-based coagulation tests. As a result, the coagulation-based hemotherapy (CBH) service was initiated by our clinical pathology team at the University of Texas at Houston Medical School in collaboration with the advanced heart failure service at Memorial Hermann Hospital. The CBH service is administered by clinical pathologists and includes preoperative, operative and postoperative consultations. Due to the complicated nature of most consult cases, many patients develop intravascular bleeding after CPB requiring blood transfusion. The CBH pathologist monitors coagulation parameters at several stages during the surgery. A comprehensive menu of coagulation testing with almost real-time results was established to cover all aspects of coagulation, which is used to provide transfusion support recommendations. Here, we describe our protocol in the CBH service that uses a comprehensive and rapid TAT menu of coagulation tests and transfusion management algorithms to prevent as well as manage bleeding during and after major cardiac surgery.

6.2.3 Establishment of Almost Real-Time Coagulation Parameters

The coagulation test menu offered by our service includes complete blood count (CBC), prothrombin time (PT), activated partial thromboplastin time (aPTT), functional fibrinogen (referred to as fibrinogen), D–Dimer, thrombin time, thromboelastography (TEG) with and without heparinase, VerifyNow-P2Y12 and VerifyNow-Aspirin assays. These tests were selected to cover all aspects of coagulation. The laboratory instruments are located in close vicinity (~ 70 ft) to the operating rooms (OR) to allow for rapid delivery of the samples and results. The TAT is additionally enhanced by hand delivery of samples from the OR to the stat laboratory for immediate testing. The laboratory testing is initiated immediately on arrival of specimens to the laboratory, with ordering by the pathologist occurring simultaneously. Based on laboratory test results, blood products are immediately released from a satellite blood bank located within the OR and the stat laboratory.

The stat laboratory has a Coulter hematology analyzer, a Stago coagulation analyzer, an Accumetrics instrument for VerifyNow and two TEG instruments. The satellite blood bank only dispenses blood products with blood group typing; screening and antibody detection are performed in the main laboratory. Two laboratory technologists are present for operating instruments, and one blood bank technologist is also present for releasing blood products. The stat laboratory and satellite blood bank were created simultaneously with the initiation of the advanced heart failure and hemotherapy services, and they are essential components of the service.

Another essential component in the success of our service was the establishment of almost real-time coagulation parameters [7]. TAT for laboratory tests for the CBH service ranged from 16 ± 10.3 min for a CBC to 20 ± 9.5 min for a coagulation panel (which includes PT, aPTT, fibrinogen, thrombin time and D-Dimer) and 27.8 ± 9.9 min for VerifyNow tests. The TAT for TEG is on average 63.7 ± 19.1 min. However, essential TEG components are available off the instrument after achieving maximum amplitude (MA) at approximately 30 min. TAT of our stat laboratory is significantly shorter than average TATs for stat coagulation results performed in central laboratories, which are in general between 45 and 60 min in typical settings. The ability to obtain laboratory results in a close to real-time manner allows the CBH pathologist to preempt the exact cause of coagulopathy as it occurs.

Assessment of bleeding risk is performed preoperatively for patients scheduled for surgery. Refer to the chapter "Preoperative Assessment of Patients" for more details. The following is a summary of preoperative evaluation. Specific risk factors for bleeding associated with cardiac surgery are elicited from the patient, including personal or family history of bleeding and history of previous cardiovascular surgery, kidney disease or liver disease. A thorough medication review for anticoagulants or antiplatelet medications with date of last use is conducted. Baseline laboratory values are obtained, including liver function test panel (alanine aminotransferase, aspartate aminotransferase, alkaline phosphatase and bilirubin), blood urea nitrogen (BUN), creatinine, CBC, PT, aPTT and fibrinogen. Abnormal results on the preoperative coagulation evaluation are further investigated by the pathologist by performing additional tests such as PT/aPTT mixing study, thrombin time, anti-Xa assay, specific factor assays and inhibitors, lupus anticoagulant or platelet aggregation studies, as clinically indicated. All results are interpreted by the CBH pathologist and communicated with the heart failure team.

6.2.4 Intraoperative Management

CPB uses an extracorporeal circuit to provide oxygenation to tissues while allowing surgery to be conducted in a bloodless, motionless field [8]. Venous cannulation with tubing removes blood from the lungs and heart to a reservoir, where it is oxygenated and returned to the cannulated arterial vasculature by a pump. Patients are anticoagulated with heparin just prior to initiation of CPB to prevent clot formation in the circuit. Once the surgery has been completed, patients are rewarmed and circulating blood is hemoconcentrated. Finally, CPB circuit is discontinued and heparin is reversed with protamine.

The CBH pathologist remains in the vicinity of the OR and monitors the patient's laboratory test results at several stages of the surgery (Table 6.1). The first set of laboratory tests is collected after induction of anesthesia as the baseline, and it includes CBC, PT, aPTT, fibrinogen, thrombin time, D-Dimer, antithrombin activity, VerifyNow-Aspirin if the patient has a history of aspirin use, and VerifyNow-P2Y12 if the

Table 6.1 Laboratory tests ordered for each phase of surgery

Test	Induction of anesthesia (baseline)	Just prior to off-pump	After heparin reversal	As needed for excessive bleeding
TEG			X[a]	X
Heparinase–TEG		X		
CBC	X	X	X	X
Coagulation panel[b]	X	X	X	X
Antithrombin	X			
VFN–Aspirin[b,c]	X			
VFN–P2Y12[b,c]	X			

[a]Optional.
[b]Coagulation panel includes prothrombin time, activated partial thromboplastin time, fibrinogen, thrombin time and D-Dimer; VFN-Aspirin, VerifyNow for Aspirin; VFN-P2Y12, VerifyNow for P2Y12.
[c]Performed if a pharmacologically induced platelet dysfunction is suspected or medication history is not known at the time of surgery.

patient is on clopidogrel, prasugrel or ticagrelor. The baseline laboratory tests may be omitted for stable patients with very recent preoperative laboratory test results. Blood products are not usually necessary while the patient is on CPB because the patient needs to be hypocoagulable during this part of the surgery. Only red blood cells (RBCs) are needed to keep the hemoglobin in the range of 8−10 g/dL during CPB. The specimen for the second set of tests is collected at the rewarming/after hemoconcentration phase of surgery, just prior to discontinuation of CPB. The specimen for a third set of laboratory tests is collected after the patient is off CPB, 10 min after heparin reversal with protamine, and before blood products are administered.

Microvascular bleeding, if present during surgery, typically occurs immediately after discontinuation of CPB. An optimal strategy for management of acute bleeding is based on rapid coagulation test results that can guide transfusion in terms of blood product type and dosage needed (Fig. 6.1).

Patients frequently develop microvascular bleeding after CPB secondary to a variety of coagulation abnormalities [6]. The laboratory parameters collected at approximately the time of CPB discontinuation are the most critical because they allow the CBH pathologist to identify the underlying cause of bleeding if it occurs—that is, bleeding secondary to

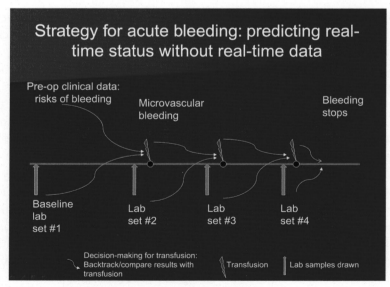

Figure 6.1 Strategy for management of acute bleeding.

coagulopathy and not due to surgical bleeding. These results are used to determine the underlying cause of coagulopathy if present, and they are used to determine the blood component types and number of units needed to transfuse to the patient after CPB discontinuation and heparin neutralization.

Due to lack of real–time laboratory results after CPB discontinuation, transfusion after heparin reversal is typically based on the laboratory results obtained prior to discontinuation of CPB (the second set of tests), with adjustment based on the postprotamine results (the third set of tests) when they become available. Selection of blood components is based on a predefined transfusion algorithm (Table 6.2).

The CBH pathologists at our institution, which consist of hemato-pathologists and transfusion medicine physicians, developed the transfusion algorithms by reviewing the literature for algorithms used during cardiac pulmonary bypass surgery and hence the algorithms are based on standard transfusion practice [9,10]. The group prepared a comprehensive hemotherapy guide, which includes all the algorithms, and the current version of this guide can be reviewed on our web site at http://hemepathreview.com. Criteria for transfusion are listed in the references section of this guide. Furthermore, the group periodically reviews the algorithms, and such algorithms are continuously revised based on experience learned on CBH service. The guide is also annually reviewed and approved by the hospital's Advanced Heart Failure Clinical Committee. Relevant criteria on medications are reviewed with input from clinical pharmacists assigned to advanced heart failure service.

The CBH pathologist remains near the OR for the duration of the surgery to select appropriate laboratory tests and blood components to order. The pathologist works with the anesthesiologist to ensure that the correct blood products as well as the correct number of units are administered to the patient to treat coagulopathy while simultaneously avoiding inappropriate use of products.

Consultations are occasionally requested when the patient is already in the OR with bleeding, usually in the context of emergency surgery. When this occurs, the clinical information is obtained from the surgical team, including the number and type of blood components already transfused. A new baseline set of coagulation tests is ordered. The hemotherapy pathologist interprets the results and manages the coagulopathy using the available data and the transfusion algorithm in the same manner as for consult cases with preoperative evaluation.

Table 6.2 Details of transfusion algorithm

Coagulation issue	Laboratory abnormality and action
Use more protamine to neutralize excess heparin	If aPTT >45 s and TT >25 s, protamine 50 mg/70 kg BW may be administered
Quantitative or qualitative platelet defect	If platelet count is >50,000/μL but <100,000/μL, then 2 single-donor apheresis units may be administered (platelet count should increase by 60,000/μL)
	If platelet count is <50,000/μL, then 3 single-donor apheresis units may be administered (platelet count should increase by 90,000/μL)
	OR
	If 35 < MA < 45 and EPL <15 and Ly30 <8, then 2 single-donor apheresis units may be administered (platelet count should increase by 60,000/μL)
	If MA <35 and EPL <15 and Ly30 <8, then 3 single-donor apheresis units may be administered (platelet count should increase by 90,000/μL)
	OR
	If VFN-P value is >130 PRU but <210 PRU, then 2 single-donor apheresis units may be administered (platelet count should increase by 60,000/μL)
	If VFN-P value is <130 PRU, then 3 single-donor apheresis units may be administered (platelet count should increase by 90,000/μL)
	OR
	If VFN-A value is >350 but <550, then 2 single-donor apheresis units may be administered (platelet count should increase by 60,000/μL)
	If VFN-A is <350, then 3 single-donor apheresis units should be administered (platelet count should increase by 90,000/μL)
Clotting factor deficiency	If 10 < hTEG R < 15, then 2 units of FFP may be administered (clotting factors should increase by 10%)
	If 15 < hTEG R < 20, then 4 units of FFP may be administered (clotting factors should increase by 20%)

(*Continued*)

Table 6.2 (Continued)

Coagulation issue	Laboratory abnormality and action
	If 20 < hTEG R, then 6 units of FFP may be administered (clotting factors should increase by 30%) If PT:20−25 or aPTT: 45−50, then 2 units of FFP may be administered If PT > 25 or aPTT > 50, then 4 units of FFP may be administered
Mild fibrinogen deficiency	If fibrinogen: 150−200 or alpha: 20−45 with normal MA, then 2 units of FFP may be administered
Marked fibrinogen deficiency or uremic platelet dysfunction	If fibrinogen <150 or chronic renal failure and bleeding with normal coagulation results, then 10 units of cryoprecipitate may be administered
Primary fibrinolysis	If EPL >15% or Ly30 >8% and MA <50 or CI <1.0, then tranexamic acid, 1000 mg/10 mL may be infused over a 10-min period OR If fibrinogen <150 and D-Dimer >10, then tranexamic acid, 1000 mg/10 mL may be infused over a 10-min period
Mild antithrombin deficiency	If antithrombin 35−50%, then 2 units of FFP prior to heparin administration is recommended
Severe antithrombin deficiency	If antithrombin <35%, then ATIII concentrate may be administered Dose (IU) = (desired level − baseline level) × weight (kg)/1.4 prior to heparin administration
Anemia	If hemoglobin <10 g/dL, then administering RBCs to keep hemoglobin at least 10 g/dL is strongly recommended
Bleeding not responding to treatment	If EPL <15% or Ly30 <8% and MA <70 or CI <3.0, consider administration of recombinant factor VIIa (15 μg/kg)

Alpha, TEG alpha in degrees; aPTT, activated partial thromboplastin time in seconds; ARU, aspirin reaction units; ATIII, antithrombin III; CI, TEG coagulation index; EPL, thromboelastography estimated platelet lysis in percentage; fibrinogen units in mg/dL; hTEG, thromboelastography with heparinase; Ly30, thromboelastography percentage clot lysis at 30 min; MA, thromboelastography maximum amplitude in mm; PLT: Platelet; PRU, P2Y12 reaction units; PT, prothrombin time in seconds; RBCs, red blood cells; TEG R, thromboelastography reaction time in minutes; TT, thrombin time in seconds; VFN-Aspirin, VerifyNow for Aspirin; VFN-P2Y12, VerifyNow-P2Y12.

6.3 MISCELLANEOUS INFORMATION ON INTRAOPERATIVE MANAGEMENT OF COAGULOPATHY

In this section, various aspects of intraoperative management of coagulopathy are discussed. The following parameters do not affect therapy in acute bleeding (considered adequate hemostasis):

- Platelet: 100,000−133,000/μL
- TEG MA: 45−50 (mm)
- Fibrinogen: 200−230 mg/dL
- TEG alpha: 45−53 (degree)
- PT: 14.7−20.0 sec
- PTT: 35.8−45.0 sec

Blood components nomenclature and expected effect on coagulation parameters on a patient with a body weight (BW) of 70 kg are as follows [9]:

- One dose of cryoprecipitate = 10 units (fibrinogen level increased by 50−100 mg/dL), 150 mL, that typically come in two half-doses (5 units each).
- One jumbo FFP = 2 single FFPs (clotting factors increased by 15−20%, fibrinogen increased by 30 mg/dL), 500 cc. Dose of FFP at 10−20 mL/kg can raise most coagulation factors levels in a nonbleeding patient by 25−50% [11].
- One unit apheresis platelet = 5 or 6 random-donor platelet (platelet increased by 30,000/μL), 200−400 mL.
- One unit RBCs: 350 mL (200 mL apheresis unit), 1 unit for 1-g increase in hemoglobin (Hgb).

When fibrinogen level is available for a stable (nonbleeding) patient, the following formula may be used to increase fibrinogen level to 100 mg/dL with cryoprecipitate:

$$\text{Units of cryoprecipitate} = (100 - \text{Fibrinogen}) \times 40$$
$$\times \text{Body weight (kg)}/25{,}000$$

However, it is important to note that the guidelines for the management of active bleeding currently indicate that the trigger level for supplementing fibrinogen should be 1.5−2.0 g/L rather than 1.0 g/L [12].

Dosing of antithrombin (AT; Thrombate), used for a very low antithrombin III (ATIII) level (<35%, which may cause heparin resistance), is based on the following formula:

$$\text{Dose (IU)} = (\text{Desired level} - \text{Baseline level}) \times \text{BW (kg)}/1.4$$

The pathologist should work with the anesthesiologist to infuse AT prior to heparinization during CPB. It is important to note that a targeted range of activated clotting time (ACT) >400 s with UFH is used during CPB (typically, 450−480 s). However, if ACT values much lower than 400 s are seen after standard dose of UFH (300 units/kg or 3 mg/kg), heparin resistance is typically the cause. This may be seen in patients with ongoing heparin intravenous or congenital AT deficiency (rare). Heparin doses in excess of 700 units/kg may not prolong ACT adequately in this case. In mild to moderate heparin resistance, additional heparin (up to three times the normal dose) may be attempted before FFP or AT concentrate is considered.

To correct Hgb for dilutional effect on Hgb due to significant FFP and platelet transfusion, the following approach may be used:

- For one unit of FFP, the patient needs one unit of RBCs (at least 0.5 on-half unit of RBCs).
- For one apheresis platelet unit, the patient needs one unit of RBCs.

(Number of RBC for Hgb correction = 0.5 × FFP + Apheresis platelets)

- For massive transfusion, it is advisable to use a 1:1:1 ratio (6 FFP, 6 RBC, 6 random-donor platelets or 1 pheresis platelet) or a 1:0.5:1 ratio (1 RBC/0.5 FFP/1 random-donor platelet or 0.2 pheresis platelets), with rounding of units.

    Minimum platelet count when the patient is on pump and off pump (on and off CPB) is as follows:
- When the patient is still on pump; platelet count needs to be kept above $20 \times 10^3 \ \mu L^{-1}$. If platelet count is already above 20,000/μL and the patient still bleeds, transfuse with platelets may be needed.
- When patient is off pump with microvascular bleeding, transfuse with platelets to keep platelet count above 100,000/μL [13].

    Quantitative D-dimer (based on immunoturbidimetric method) may not be accurate (up to 15% discrepancy) if the patient has marked hemolysis. Semiquantitative fibrin/fibrinogen split product (FSP) may be used instead if needed.

    For targeted ACT with heparin reversal using protamine, the anesthesiologist typically uses protamine to reverse heparin after CPB with targeted ACT of less than 130 s (or within 10% of pre-CPB ACT).

    It is important to note that 1 mg of protamine is used to neutralize 1 mg (100 units) of UFH. For a patient with a BW of 70 kg (which requires 21,000 units of UFH with 300 units/kg dosage), it is approximately 210 mg protamine. Some anesthesiologists may use 1.3 mg of protamine to neutralize 1 mg of UFH (273 mg protamine for 70-kg BW).

For a high dose of heparin in CPB (300–400 IU/kg), the half-life of heparin is 126 ± 24 min [14]. Plasma heparin concentration is 3 or 4 IU/mL. Hypothermia delays heparin elimination [15]. Chronic renal failure also prolongs the elimination of heparin [16]. However, liver disease has no effect on heparin elimination [17].

At the end of CPB, patients may have adequate hemostasis with normal ACT after heparin neutralization, but they may develop bleeding several hours later with prolonged clotting time. This phenomenon, known as heparin rebound [18,19], results from delayed release of heparin that is previously sequestered in tissues (especially in adipose tissue in obese patients) into the circulation. In this case, additional protamine is given to neutralize heparin.

It is important to note that a prolonged ACT may be seen in patients without residual heparin. It may be due to other etiologies (hemodilution, coagulation factor depletion, thrombocytopenia, platelet dysfunction, hypothermia and even excess protamine) [20]. Therefore, more sensitive tests (PTT and TT) should be used to assess residual heparin that may benefit from additional protamine.

Falsely elevated Ly30 value in TEG may occasionally be seen with kaolin as activator in TEG for citrated blood [21]. Other lab data, such as normal D-dimer, FSP and fibrinogen, may be assessed to rule out actual fibrinolysis.

Tranexamic acid for hyperfibrinolysis:

- Hyperfibrinolysis may accompany cardiac surgery using CPB, in patients with liver disease [22], resulting in microvascular bleeding [23]. Clinical suspicion should be high in cases in which bleeding continues despite hemostatic replacement therapy. In such cases, platelet levels are relatively conserved but fibrinogen levels are disproportionately low, and D-dimer levels are disproportionately high for DIC in stage II. TEG Ly30, which may help differentiate fibrinolytic activation from coagulation factor deficiency, is not very sensitive because it detects only the most marked changes [24].
- Tranexamic acid is used to inhibit hyperfibrinolysis. It is 10 times more potent than Amicar (EACA).
- Tranexamic acid is contraindicated in patients with evidence of DIC in stage 1 (hypercoagulation state with secondary fibrinolysis) because

the fibrinolytic system is required to ensure the dissolution of the widespread fibrin [22,25].

Platelet dysfunction in uremia:

- Almost all patients with uremia, the clinical syndrome of advanced renal failure, have a bleeding diathesis [26]. This predisposition becomes especially problematic when these patients undergo invasive procedures such as surgery, biopsy or catheter placement. Moreover, many of the clinical presentations of uremic bleeding involve life-threatening conditions, including pericardial tamponade, intracranial bleeding and gastrointestinal bleeding.

- In hemodynamically unstable patients with uremia, the massive occult bleeding that can occur in these conditions is particularly troubling and should remain a central concern in the evaluation of such patients. When the BUN level is greater than 60 mg/dL or the serum creatinine level is greater than 6.7 mg/dL, bleeding time is significantly prolonged [27].

- Bleeding time is currently not considered a reliable test because it has a high incidence of false positivity and false negativity. Platelet aggregation study may show a decrease in platelet response to various reagents, but the threshold for bleeding risk has not been established. Without established criteria, we use an empirical threshold of BUN greater than 45 for cryoprecipitate transfusion in a bleeding patient.

6.4 POSTOPERATIVE ASSESSMENT AND MANAGEMENT OF COAGULOPATHY

Patients who undergo cardiovascular (CV) surgeries may have a high risk of postoperative bleeding due to the following factors: long pump time due to complicated surgeries, redo (resternotomy due to previous CV surgery), renal disease, liver disease, residual anticoagulant/antiplatelet medications, obesity (risk of heparin rebound) and massive transfusion in surgery causing right ventricular failure. Cardiac tamponade due to bleeding in the thoracic cavity can impair cardiac function (including Total Artificial Heart with compression of pulmonary vessels), and the resulting stasis can lead to increasing chest tube output.

Mortality from acute blood loss may be eminent, requiring massive transfusion in many of these patients. Severe coagulopathy is associated with a less favorable outcome that is statistically significant when late mortality is included. This late mortality is almost always caused by multiorgan failure secondary to prolonged hypotension and hypoperfusion of major organs at the time of the acute event. The etiology of coagulopathy is probably multifactorial. As any episode progresses, dilutional effects are difficult to differentiate from those associated with DIC. DIC may be triggered by hypoperfusion-induced tissue hypoxia or endotoxin release from either necrotic tissues or transfused products. The fact that many patients involved in massive transfusion incidents develop severe coagulopathy may reflect not only the underlying etiology and the severity of the hemorrhage but also a failure to adequately replace coagulation factors. Our key strategy for treatment for massive hemorrhage includes timely blood component transfusion to normalize coagulation parameters using real-time stat laboratory data for rapid transfusion.

During the postoperative period in the cardiovascular intensive care setting at our institution, a standard postoperative order by a surgeon would look similar to the following:

- Physician to nurse order: If output from one chest tube (CT) >150 mL/h, send for CBC and platelet count, DIC screen, call cardiologists, surgeon and hemotherapy.
- When there is significant microvascular bleeding (defined as 150–200 mL/h or more from CT drain for several hours without obvious clinical evidence of surgical bleeding), the patient's attending (cardiologist or surgeon) would request for hemotherapy consultation via the blood bank or call the hemotherapy pathologist directly.
- If the request is received by the blood bank, technologists would obtain the contact information from the requesting clinicians. The pathologists on call (for hemotherapy consultation) will be notified. The pathologist will obtain clinical information from this contact. The intensive care unit (ICU) nurse who takes care of the patient can be contacted through the charge nurse for information on the most recent coagulation results, most recent transfusion and current transfusion. Further detailed clinical information is also available from electronic medical records.
- Transfusion is delegated to Hemotherapy Service, and the pathologist will order blood components as dictated by the laboratory results and clinical findings. Cases with significant bleeding typically need on-site presence of pathologists. Issues related to transfusion need to be discussed with the patient's attending (surgeon in the immediate perioperative period or cardiologist otherwise). Igloos containing blood components left over from surgery are typically taken to the patient's

ICU room for potential use. We typically check with patient's nurse to determine if there are enough units to transfuse; otherwise, more blood products are ordered from the blood bank.

- For significant bleeding without baseline, the following protocol is used:
  - Order stat baseline CBC and platelet count as well as DIC panel.
  - If CT output is significant, the following rules–of–thumb can be used to start replacing blood loss: For every 500 cc of CT output, give 1 RBC, 0.5 FFP and 0.2 platelet (using 1/0.5/1 ratio; ie, 1 RBC/0.5 FFP/1 random-donor platelet, with rounding). Refer to Table 6.3 for units of blood components to replace blood loss. For example, 2000 cc blood loss would need 4 RBCs, 2 FFP and 1 unit of platelet for replacement. Any further correction for coagulopathy would require additional blood components (eg, more platelets for thrombocytopenia).
  - When laboratory results are available (in ~30 min), adjust transfusion component type/dose.
  - Subsequent sets of tests may need to be ordered to evaluate treatment.
  - Hematocrit can also be done in the patient's room by the nurse with Gem Premier Instrument (which also runs blood gases) if available on-site.

Once coagulation parameters have been almost restored, acute bleeding should subside. If bleeding continues or gets worse in this situation, consider anatomical etiology (surgical bleeding such as due to anastomosis leakage) and discuss the use with cardiovascular surgeons. Cardiovascular surgeons are typically informed by the patient's nurse in significant bleeding situations and usually stand by for possible emergency reoperation. Even if bleeding is caused by a surgical source, normalizing coagulation parameters is helpful to prepare the patient for emergently exploratory open–chest surgery.

Table 6.3 Replacement of blood loss

Blood loss (CT output) (cc)	RBC (units)	FFP (units)	Platelet (apheresis units)	Total blood component volume (cc)[a]
500	1			300
1000	2	1		850
1500	3	1		1150
2000	4	2	1	2000

[a]1 RBCs 300 cc, 1 FFP 250 cc, 1 PLT 300 cc.

For bleeding patients with elevated INR, after transfusion of multiple FFP units (typically defined as four or more), if INR is still elevated (with adequate platelet count/function and fibrinogen), instead of transfusing more FFP, use KCentra to avoid volume overload. Refer to the chapter "Antiplatelets and Anticoagulants" for details.

If CT output is more than 1000 mL/h and not alleviated after significant transfusion, administration of recombinant factor VIIa (rFVIIa) should be considered. It is important to be aware of complications due to the use of rFVIIa and its limitation in patients' outcome and also the 11% rate of thrombotic complications. Refer to the chapter "Antiplatelets and Anticoagulants" for details.

Once bleeding has improved (<150 mL/h with decreasing trend), usually no further treatment is needed. Another set of lab tests can be ordered in 2 h with instructions for the patient's nurse to call pathologists if bleeding worsens or lab results are critical. Less than 150 mL/h is the targeted level for acute bleeding. Chest tubes can be removed once drain is less than 30–50 mL/h. Anticoagulant then can be started.

Without acute bleeding in postoperative period, the following are typical threshold for transfusion:

- Hgb <8 g/dL, fibrinogen <150 mg/dL and platelet < 50,000/μL: Standing orders for transfusion can be ordered to direct transfusion (see the section "Difficult Problems in Coagulopathy Management" in the chapter "Specific Clinical Situations" for patients who are transfusion-dependent).
- Consideration for central venous pressure (CVP) in transfusion: CVP is a good approximation of right atrial pressure, which is a major determinant of right ventricular end diastolic volume. CVP is used as a surrogate for preload. Normal values are 5–10 cm H_2O. Transfusion would cause volume overload if CVP >20 cm H_2O. Transfusion would need to be coordinated with the patient's cardiologist in this case. CVP can be decreased by hemodilution (blood volume taken out if patient is on continuous venovenous hemodialysis, provided that mean arterial pressure can be kept >60 cm H_2O), or giving patient Bumex (diuretic, goal CVP <14 cm H_2O) or adjusting vacuum pressure in LVAD/ BiVAD (patients on assist devices). Goals are typically set for <11–13, 15 or 20 cm H_2O, depending on cardiac status.
- Consideration for mean arterial pressure (MAP) in transfusion: MAP is the average arterial pressure during a single cardiac cycle. A MAP that is greater than 60 mm Hg is enough to sustain the organs of the

average person. MAP is normally between 70 and 110 mm Hg. Note that in the postoperative period, surgeons may order RBCs (together with FFP and platelets to avoid dilutional effect) to keep MAP at desired level. Such transfusions would need to be factored into transfusion to correct coagulopathy. Goals are typically set for MAP >55 or 60 mm Hg, depending on cardiac status.

6.5 CONCLUSIONS

This chapter discussed one of the most critical components of the hemotherapy service—assessment and management of perioperative coagulopathy. An emphasis was placed on patients undergoing surgery with CPB in which significant alterations in hemostasis occur during and after surgery. Strategies to optimize coagulopathy management were outlined on two key areas: (1) stat testing setup to obtain almost real-time laboratory results with rapid turnaround time and (2) timely therapy with blood components or pharmaceutical products based on well-defined algorithms.

REFERENCES

[1] Massad MG, Cook DJ, Schmitt SK, Smedira NG, et al. Factors influencing HLA sensitization in implantable LVAD recipients. Ann Thorac Surg 1997;64:1120—5.
[2] McKenna Jr. DH, Eastlund T, Segall M, Nooren HJ, et al. HLA alloimmunization in patients requiring ventricular assist device support. J Heart Lung Transplant 2002;21:1218—24.
[3] Moazami N, Itescu S, Williams MR, Argenziano M, et al. Platelet transfusions are associated with the development of anti-major histocompatibility complex class I antibodies in patients with left ventricular assist support. J Heart Lung Transplant 1998;17:876—80.
[4] Goldstein DJ, Beauford RB. Left ventricular assist devices and bleeding: adding insult to injury. Ann Thorac Surg 2003;75:S42—7.
[5] Drakos SG, Stringham JC, Long JW, Gilbert EM, et al. Prevalence and risks of allosensitization in HeartMate left ventricular assist device recipients: the impact of leukofiltered cellular blood product transfusions. J Thorac Cardiovasc Surg 2007;133:1612—19.
[6] Whitlock R, Crowther MA, Ng HJ. Bleeding in cardiac surgery: its prevention and treatment—an evidence-based review. Crit Care Clin 2005;21:589—610.
[7] Welsh KJ, Nedelcu E, Bai Y, Wahed A, et al. How do we manage cardiopulmonary bypass coagulopathy? Transfusion 2014;59:2158—66.
[8] Machin DA, Allsager C. Principles of cardiopulmonary bypass. Contin Educ Anaesth Crit Care Pain 2006;6:176—81.
[9] American Red Cross. Practice guidelines for blood transfusion: a compilation from recent peer-reviewed literature. 2nd ed. http://www.sld.cu/galerias/pdf/sitios/anestesiologia/practical_guidelines_blood_transfusion.pdf/; 2007 [accessed 28.09.15].

[10] Levy JH, Tanaka KA. Management of surgical hemostasis: systemic agents. Vascular 2008;(Suppl. 1):S14–21.

[11] Sussman IL, Spivack M. Indications and use of fresh frozen plasma, cryoprecipitate, and individual coagulation factors. In: Dutcher JP, editor. Modern transfusion therapy I. Boca Raton, FL: CRC Press, Inc; 1990. p. 209–10.

[12] Spahn DR, Cerny V, Coats TJ, Duranteau J, et al. Management of bleeding following major trauma: a European guideline. Crit Care 2007;11:R17 [Erratum, Crit Care 2007;11:414].

[13] College of American Pathologists. Practice parameter for the use of fresh frozen plasma, cryoprecipitate and platelets. JAMA 1994;271:777–81.

[14] Olsson P, Lagergren H, Ek S. The elimination from plasma of intravenous heparin. An experimental study on dogs and humans. Acta Med Scand 1963;173:619–30.

[15] Cohen JA, Frederickson EL, Kaplan JA. Plasma heparin activity and antagonism during cardiopulmonary bypass with hypothermia. Anesth Analg 1977;56:564–70.

[16] Cipolle RJ, Seifert RD, Neilan BA, Zaske D, et al. Heparin kinetics: variables related to disposition and dosage. Clin Pharmacol Ther 1981;29:387.

[17] Sette H., Hughes R.D., Langley P.G., Gimson A.E. et al. Heparin response and clearance in acute and chronic liver disease. Thromb Haemost 1985;54:591.

[18] Kaul TK, Crow MJ, Rajah SM, Deverall PB, et al. Heparin administration during extracorporeal circulation. J Thorac Cardiovasc Surg 1979;78:95.

[19] Perkin HA, Osborn JJ, Gebode F. The management of abnormal bleeding following extracorporeal circulation. Ann Intern Med 1959;51:658.

[20] Turman KJ, McCarthy RJ, Djuric M, Rizzo V, et al. Evaluation of coagulation during cardiopulmonary bypass with a heparinase-modified thromboelastographic assay. J Cardiothorac Vasc Anesth 1994;8:144–9.

[21] Durila M. Kaolin activated thromboelastography can result in false positive fibrinolytic trace. Letter to the editor. Anaesth Intensive Care 2011 July 1.

[22] Hunt BJ. Bleeding and coagulopathies in critical care. N Engl J Med 2014;370:847–59.

[23] Attar S. Hemostasis in cardiac surgery. New York: Futura Publishing Co; 1999. p. 205.

[24] Raza I, Davenport R, Rourke C, Platton S, et al. The incidence and magnitude of fibrinolytic activation in trauma patients. J Thromb Haemost 2013;11:307–14.

[25] Physician desk reference. Montvale, NJ: Medical Economics Data Production Co; 1995. p. 1157.

[26] Sinasi Salman S. Uremic bleeding: pathophysiology, diagnosis, and management. Hosp Physician May 2001;45–76.

[27] Huang R, Nedelcu E, Bai Y, Wahed A, et al. Postoperative bleeding risk stratification in cardiac pulmonary bypass patients using artificial neural network. Ann Clin Lab Sci 2015;45(2):181–6.

CHAPTER 7

Specific Clinical Situations

Contents

A. Nguyen, A. Dasgupta, A. Wahed:
Management of Hemostasis and Coagulopathies for Surgical and Critically Ill Patients. © 2016 Elsevier Inc.
DOI: http://dx.doi.org/10.1016/B978-0-12-803531-3.00007-0   All rights reserved.

7.1 INTRODUCTION

Evidence-based hemotherapy is practiced in specific clinical situations, especially in the field of cardiology and cardiac surgery to effectively manage bleeding issues of critically ill patients. There are numerous such scenarios, including mechanical assist devices [eg, left ventricular assist device (LVAD), artificial heart and extracorporeal membrane oxygenation (ECMO)], heparin-induced thrombocytopenia (HIT)-positive patients undergoing cardiac surgery, lupus antibody-positive patients on anticoagulants monitoring and other situations. These topics are divided into two major categories: (1) typical anticoagulant and antiplatelet protocols for specific situations and (2) difficult problems in coagulopathy management.

7.2 TYPICAL ANTICOAGULANT AND ANTIPLATELET PROTOCOLS FOR SPECIFIC SITUATIONS

In this category, various clinical situations are discussed, including patients with LVAD and various other clinical situations, such as tandem heart system, intra-aortic balloon pump (IABP), care of patients after undergoing coronary artery bypass grafting (CABG), situations with percutaneous coronary intervention, patents on ECMO, anticoagulation and antiplatelet therapy in patients receiving total artificial heart (SynCardia) and anticoagulation following bioprosthetic valve implantation. Moreover, cardiopulmonary bypass anticoagulation with unfractionated heparin (UFH) or bivalirudin (Angiomax) for patient with HIT is also discussed in this section.

7.2.1 Patients With Left Ventricular Assist Device

Patients with continuous-flow devices require anticoagulation and antiplatelet agents to attenuate the risk of thromboembolic events. Various anticoagulant protocols have been developed since the HeartMate II Pivotal Trial [1]. At our institution, patients are typically put on acetylsalicylate (ASA; aspirin) 81—325 mg daily, UFH to achieve a target partial thromboplastin time (PTT) range of 60—80 s and/or Coumadin (warfarin) to achieve international normalized ratio (INR) between 2.5 and 3.5. Heparin may be discontinued after overlap with Coumadin and achieving the target INR for 2 or 3 consecutive days.

ASA can be typically started on the second postoperative day unless chest tube (CT) output is high. However, when CT drainage reaches

<50 mL/h along with no evidence of bleeding, removal of chest tubes may be performed and intravenous heparin may be started on the first postoperative day. Patients typically receive fixed heparin dose. The following are typical orders:

- Heparin infusion may be initiated at a particular time with a fixed dose of 200 units/h and checking PTT every 6 h. Physician should be notified if PTT is >60 s.
- Heparin infusion not intended to achieve therapeutic PTT >60 s.
- No heparin loading dose is used.

If therapy is initiated with a fixed heparin dose, then such dosage may be slowly increased (eg, to 400 units/h on the second postsurgical day to 800 units/h on the third postsurgical day). Once the patient is doing well without bleeding, with initial starting dose on the third postoperative day, then order is placed for "Heparin Weight-Based Acute Coronary Syndrome protocol" (therapeutic PTT 60–80 s, no loading dose). Refer to the chapter "Antiplatelets and Anticoagulants" for more details.

If clots develop in a patient with LVAD, it is preferable to use bivalirudin (rather than UFH) as anticoagulant. Plasma levels of lactate dehydrogenase (LDH) and free plasma hemoglobin may be significantly elevated due to clot in a patient LVAD. Clot may lead to LVAD failure, manifested by an insignificant increase in left ventricular ejection fraction with increase in speed (in rpm) or power (in watts). This may require LVAD exchange.

7.2.2 TandemHeart System

The TandemHeart system requires that the patient must be anticoagulated. It is recommended that the activated clotting time (ACT) should be 400 s for insertion of the device in the catheterization lab or operating room. The ACT should be >200 s during the support period. If ACT is unavailable, activated partial thromboplastin time (aPTT) can be used. During support, aPTT should be maintained between 55 and 75 s. The TandemHeart system will deliver 900 units/h at the recommended dosage of 90 units/cc of heparin (10 cc/h) to the patient through the infusation system of the pump. Additional heparin may be required to be administered peripherally to maintain proper anticoagulation levels. See Table 7.1 for a typical UFH protocol for TandemHeart.

A typical UFH infusion rate for TandemHeart is 900 units/h. For patients with low body weight (eg, 50 kg), an average UFH rate of

Table 7.1 A typical UFH protocol for TandemHeart with therapeutic PTT range of 55–75 s

PTT range (s)	Action
<55	Continue current concentration of device heparin and initiate systemic heparin at 2 units/kg/h; recheck PTT in 2 h.
55–75	Therapeutic—no change.
76–90	Switch device heparin to 25,000 units in NS 500 mL at 500 units/h (10 mL/h) and initiate systemic heparin at 2 units/kg/h; recheck PTT in 2 h.
91–110	Switch device heparin to 25,000 units in NS 500 mL at 500 units/h (10 mL/h); recheck PTT in 2 h.
>110	Switch device heparin to NS at 10 mL/h; recheck PTT in 1 h.

900 units/h may cause PTT to be in the supratherapeutic range. In that case, the TandemHeart heparin device is set to run normal saline (NS), and the patient is put on a weight-based UFH protocol, such as infusion protocol for heparin weight-based acute coronary syndrome (ACS) (therapeutic PTT of 60–80 s).

Prior to cardiovascular surgery, heparin is only stopped when the patient is ready to leave the intensive care unit (ICU) room for the operating room. Typically, actual surgery does not start until 2 or 3 h later. Patients may still have prolonged PTT, but this does not present risks to the patient.

7.2.3 Intra-Aortic Balloon Pump (IABP)

The existing data suggest that it is safe to omit heparinization when using IABP counterpulsation [2]. The decision to heparinize should be weighed in the context of other indications or contraindications rather than being an automatic response to the use of IABP. Most cardiologists, including those at our institution, prefer to use heparin with a targeted range for PTT of 60–80 s. Prior to cardiovascular surgery, heparin is only stopped when the patient is ready to leave the ICU room for the operating room. Typically, actual surgery does not start until 2 or 3 h later. The patient may still have prolonged PTT, but this does not present risks to the patient.

7.2.4 Post Coronary Artery Bypass Graft

The Seventh ACCP (American College of Chest Physicians) Conference on Antithrombotic and Thrombolytic Therapy made the following recommendations for the prevention of saphenous vein graft occlusion following CABG, also known as aortocoronary bypass (ACB) [3]:

- Aspirin (75−162 mg/day) treatment for indefinite period of time for all patients with coronary artery disease is recommended.
- Postoperative aspirin (75−162 mg/day) starting 6 h after CABG procedure preferred over preoperative aspirin.
- Addition of dipyridamole to aspirin therapy is not recommended in patients undergoing CABG.
- For coronary artery disease patients undergoing CABG who are allergic to aspirin, clopidogrel 300 mg loading dose 6 h postoperatively followed by 75 mg/day maintenance dosage is recommended.
- Clopidogrel (75 mg/day) in addition to aspirin for 9−12 months following procedure in patients undergoing CABG for non-ST segment elevation acute coronary syndrome is recommended.
- Aspirin is recommended for all patients after CABG within 24 h with adequate hemostasis. It is not recommended to give aspirin to patients after CABG if they are already on Coumadin (eg, for patients with a history of chronic atrial fibrillation). There is no evidence that it improves graft patency, and it certainly increases the risk for bleeding. On the other hand, there is evidence that the addition of aspirin to Coumadin in patients with mechanical valves results in a further decrease in thromboembolic complications after mechanical valve replacement. There is no evidence to support recommending administration of clopidogrel to all patients after CABG.

7.2.5 Percutaneous Coronary Intervention

The Seventh ACCP Conference on Antithrombotic and Thrombolytic Therapy made the following recommendations for patients undergoing percutaneous coronary intervention (PCI):

- Aspirin pretreatment (75−325 mg), followed by aspirin 75−162 mg/day for long-term treatment after PCI is recommended.
- Lower doses of aspirin (75−100 mg/day) for long-term treatment in patients receiving other antithrombotic/anticoagulant agents, such as clopidogrel or Coumadin, is recommended.
- In patients who have undergone stent placement, combination therapy with aspirin and thienopyridine derivative (clopidogrel) is preferred over systemic anticoagulation therapy.

- Clopidogrel (75 mg/day) in addition to aspirin for at least 9−12 months after PCI is recommended. It is interesting to note that patients who undergo CABG soon after PCI only need ASA after surgery (except for those with non-ST segment elevated myocardial infarction).
- For patients with low atherosclerotic risk (eg, isolated coronary lesion), clopidogrel for at least 2 weeks after bare metal stent placement, for 2 or 3 months after placement of a sirolimus-eluting stent and for 6 months after placement of paclitaxel-eluting stent is recommended.
- Therapy with glycoprotein (GpIIb/IIIa) antagonist such as abciximab or eptifibatide for all patients undergoing PCI, particularly those undergoing primary PCI or those with refractory unstable angina or other high-risk features, is recommended.
- Abciximab should be administered as 0.25 mg/kg bolus followed by 12-h infusion at a rate of 10 μg/min.
- Eptifibatide should be administered as double bolus (each 180 μg/km, 10 min apart) followed by 18-h infusion of 2 μg/kg/min.
- Tirofiban not recommended as an alternative to abciximab.
- For patients with non-ST segment elevation, myocardial infarction/ unstable angina (NSTEMI/UA) rated moderate to high risk based on Thrombolysis in Myocardial Infarction (TIMI) score, upstream use of GpIIb/IIIa antagonist (eptifibatide or tirofiban) should be started as soon as possible prior to PCI.
- In NSTEMI/UA patients with elevated troponin level, initiation of abciximab within 24 h prior to PCI is recommended.
- In patients not receiving a GpIIb/IIIa inhibitor, a weight-adjusted heparin bolus of 60−100 IU/kg should be administered in doses to produce an ACT of 250−350 s.
- For uncomplicated PCI, routine post-procedural heparin infusion is not recommended.
- For patients who do not receive GpIIb/IIIa antagonist, bivalirudin (0.75 mg/kg bolus followed by infusion of 1.75 mg/kg/h for duration of PCI) is preferred over heparin during PCI.
- Routine administration of Coumadin or other vitamin K antagonists is not recommended after PCI in patients with no other indications for systemic anticoagulation therapy.

7.2.6 Extracorporeal Membrane Oxygenation

Although the ECMO circuit has an anticoagulant lining, low-dose heparin is usually administered to prevent clot formation. The lowest effective level of anticoagulation is not known, and heparin may be avoided altogether if the risks of heparin therapy are considered excessive in patients with significant coagulopathy. Some patients with severe hemorrhage have safely undergone several days of ECMO without any systemic anticoagulation, although in this situation it would be advisable to avoid prolonged periods of low ECMO flow rates (<2l pm).

For venous—arterial ECMO after cardiopulmonary bypass, excessive bleeding due to coagulopathy should be managed as usual. Use of recombinant factor VIIa (rFVIIa) in patients on ECMO may be associated with acute generalized intravascular thrombosis.

Following cardiac surgery, heparin is commenced when the chest tube drainage is <100 mL/h for 2 or 3 h, the patient is normothermic, and coagulation parameters are acceptable. Heparin should ideally be commenced within 24 h postoperatively, and this is usually possible within 12 h. The dose is titrated to maintain an ACT of 150—180 s, which should be measured every 2 h until it reaches a stable level (Table 7.2). Heparin resistance is usually due to antithrombin (AT) deficiency; this may be treated with fresh frozen plasma or AT concentrate. Tranexamic acid may be infused while the patient is on ECMO to treat hyperfibrinolysis.

For venovenous ECMO, heparin infusion is commenced at 12 units/kg/h once the postcannulation kaolin ACT has fallen below 200 s. Kaolin ACT should be measured every 2 h for the first 24 h, and heparin infusion should be adjusted as needed. The target kaolin ACT in patients with platelets >80,000/μL is 150—180 s. The target should be decreased

Table 7.2 UFH monitor with activated clotting time (ACT) for patients on ECMO

ACT (s)	Response
<130	Bolus 1000 units and increase infusion 200 units/h.
130—150	Increase infusion 100 units/h.
150—180	No change.
180—200	Decrease infusion 100 units/h.
200—250	Decrease infusion 200 units/h.
>250	Cease infusion for 1 h. Check ACT hourly and recommence when ACT <200 s at 300 units/h less than the original rate.

if the patients have a tendency toward bleeding. Anticoagulation in patients with marked thrombocytopenia (platelet <20,000/μL) should be discussed with the ICU physician; in general, heparin may be ceased in these patients. In most instances, thrombocytopenia prolongs the ACT; hence, this may still be a suitable monitor of anticoagulation in mild to moderate thrombocytopenia.

After 24 h, aPTT can be used to monitor anticoagulation (target range of 55−75 s). The aPTT and ACT should be checked every 6 h, and heparin dose should be adjusted to the aPTT as per the hospital protocol. Platelets are continuously consumed because of the exposure to the foreign surface and the sheer force. As a result, platelet counts should also be monitored frequently. Fibrinogen and D-dimer should be checked daily. If the patient develops DIC with hypocoagulopathy, ECMO should be operated without anticoagulant. Severe DIC should be alleviated with blood components [red blood cells (RBCs), fresh frozen plasma (FFP), cryoprecipitate and platelet]. During transfusion, ACT or aPTT is monitored very closely by ECMO perfusionists, and heparin may be added if needed to prevent clot.

For HIT patients, direct thrombin inhibitor Bivalirudin is administered [4] with a bolus of 0.5 mg/kg followed by a continuous infusion of 0.5 mg/kg/h. Using this protocol, ACT values ranging from 180 to 220 s can be achieved.

7.2.7 Total Artificial Heart (SynCardia): Anticoagulation and Antiplatelet Therapy

After total artificial heart (TAH) implantation, anticoagulant/antiplatelet therapy is ideally started after chest closure/washout, typically on the first postoperative day, but it can be delayed further if necessary [5]. The main goal for hemostasis during follow-up is to ensure that antiplatelet medications and anticoagulants are in therapeutic ranges so that bleeding or thrombosis can be avoided.

7.2.7.1 Postoperative Period (Immediate)

In the immediate postoperative period when chest tube drainage <100 mL/h for 2 h and platelet count is <50,000/μL, the following protocols may be followed:

- Dipyridamole (Persantine) is started at 100 mg either orally (p.o.) or through nasogastric tube (NG) feeding every 8 h (75 mg for patients with body weight <70 kg).

- Administration of ASA at 81 mg p.o. or NG tube per day may be initiated; ASA may be put on hold if postoperative bleeding has been significant.
- Pentoxifylline (Trental) 200 mg is started p.o. or NG tube (oral suspension) every 8 h (400 mg if fibrinogen increased above normal). Pentoxifylline oral solution may be compounded by pharmacy for feeding tube administration if patient is intubated and cannot swallow medications (nonformulary pentoxifylline 20 mg/mL, oral suspension, 200 mg NG three times per day).
- Dipyridamole and ASA doses need to be adjusted based on platelet aggregation results. Platelet aggregation is to be performed twice per week (eg, Monday and Thursday). The first one is done 2 days after starting the medications. The goal is to obtain a decrease in aggregation ($<40\%$) with the following conditions: arachidonic acid (two concentrations), ADP (at 2.5 μm/mL only) and epinephrine (two concentrations). If aggregation with collagen is decreased ($<20\%$), this indicates that too much dipyridamole or ASA is being given. As a result, daily dosages of one or both medications should be decreased to prevent bleeding.
- Maximum dipyridamole dosage is 400 mg every 8 h. Maximum ASA dosage is 325 mg each day.
- If platelet count drops ($<50,000/\mu$L), requiring transfusion, or if patient develops acute bleeding, antiplatelet medications may need to be temporarily stopped or decreased to minimal doses (depending on the degree of bleeding) until problems are resolved.
- After antiplatelet medications are in therapeutic range with two consecutive platelet aggregation studies, testing can be spaced out to conserve blood.

Antiplatelet medications for patients with thrombocytopenia who receive TAH:

- If platelet count is $<50,000/\mu$L, platelet aggregation test should not be performed (results will be abnormal due to low count; patients are also at risk for bleeding). It is advised to keep dosages of ASA and Persantine minimal.
- If platelet count is 50,000$-$100,000/μL, use normal control with similar platelet count to adjust medications. If control percentage aggregation is normal at such platelet count, use TAH protocol with the platelet aggregation percentage. If control percentage is low and the patient's results are expected to also be abnormal, keep antiplatelet medications at minimal doses.
- If platelet count is $>100,000/\mu$L, platelet aggregation test should be performed as usual to adjust medications.

7.2.7.2 Postoperative (Chest Tubes Pulled)

Postoperatively when chest tubes are pulled and when chest tube drainage is less than 30 mL/h for at least 4 h, the following protocols are applicable:

- Heparin is typically administered and monitored in a manner similar to that for LVAD (see Section 7.2.1), with additional use of thromboelastography (TEG) results as needed.
- Optional use of TEG data: Adjustment for heparin should prevent hypercoagulation (to achieve normocoaguability; ie, CI <3.0). In general, TEG typically shows high maximum amplitude even with adequate antiplatelet medications by platelet aggregation study. The overall goal is to keep CI <3.
- If the patient develops acute bleeding on heparin, heparin infusion should be stopped until bleeding resolves.
- Duration: Heparin is given for 2 weeks and then (based on clinical status) may be switched to Coumadin to maintain INR 2.5–3.5 for 2 or 3 consecutive days, then IV heparin is stopped. If the patient is fed through an NG tube, this may interfere with absorption of Coumadin if also given through NG tube [6, 7]. It is important to flush the feeding tube following Coumadin to minimize interaction with the tube.
- Thrombin time is typically normal or only slightly prolonged even when the patient is on heparin with therapeutic range for PTT.

7.2.7.3 Transfusion Threshold (Nonbleeding Patients)

In general, if hemoglobin is less than <7 g/dL and/or platelet count is less than 50,000/µL, transfusion should be considered. It is important to monitor laboratory values daily, and tests such as CBC, disseminated intravascular coagulation (DIC) panel, lactate dehydrogenase, hematoglobulin, plasma-free hemoglobin and antithrombin level should be ordered daily.

7.2.8 Cardiopulmonary Bypass and Anticoagulation With UFH for Patients With Heparin-Induced Thrombocytopenia

Strategies for choosing perioperative anticoagulation in patients with a recent history of HIT awaiting cardiopulmonary bypass (CPB) are discussed here [8–10]. One important option is the use of plasmapheresis for patients with subacute HIT [ie, those patients with recent HIT in whom the platelet count has recovered but in whom anti-PF4/heparin Ig antibodies (HPF4) remain detectable with or without a positive heparin-induced platelet aggregation (HIPA) or serotonin release assay (SRA)]. Patients with a history of subacute HIT who require cardiovascular surgery have been successfully anticoagulated with a brief course of unfractionated heparin during CPB without complications. This approach is based on the theory that a secondary immune response after re-exposure to heparin is unlikely to occur until at least 3 days later. Thus, a brief exposure to heparin during CPB should not immediately elicit HIT antibodies. Furthermore, because heparin is rapidly cleared after the procedure with protamine neutralization, even if antibodies appeared a few days later, they would not be thrombogenic in the absence of heparin.

For patients with existing HPF4 antibody, plasmapheresis has been reported as a rescue therapy to effectively remove the antibody, thus decreasing the risk of thrombotic complications during heparin re-exposure. However, there are some general concepts. First, a direct relationship has been noted between anti-HPF4 concentrations measured by either enzyme-linked immunosorbent assay (ELISA) values or by percentage release of radioactive serotonin via the SRA and the propensity to develop thrombotic complications. Second, other antibody-mediated diseases improve with plasmapheresis, such as thrombotic thrombocytopenic purpura and myasthenia gravis; these generally start to respond when antibody levels are still detectable but reduced by 60−80%, which can be achieved from one plasmapheresis with 1−1.5 plasma volume. Third, normalization of the platelet count has been accomplished after a series of three apheresis procedures of 3-L plasma exchanges each from most of the cases reported in the literature [8−10]. There was also a significant decrease in the heparin-induced platelet aggregation between pre- and post-apheresis patient serum samples from those studies. We use the following strategy to manage patients with a history of HIT (see also Table 7.3):

- Patients with a history of HIT and a negative HPF4 screen just before surgery can safely be given heparin during CPB. However, for patients with circulating anti-HPF4 and no evidence of active HIT (platelet count already recovered), HIPA, a functional test for HPF4 antibody, should be performed.
- If anti-HPF4 is borderline positive (optical density (OD) ≤ 0.6) and HIPA is negative, patients can safely be given heparin during CPB. HPF4 level should be checked daily for 3 days after surgery.
- If anti-HPF4 is positive (OD > 0.6) and HIPA is negative, patients may still have risk for intraoperative or postoperative thrombotic

Table 7.3 Management of anticoagulant during and after surgery in patients with HIT

HPF4 (heparin-associated antibody by ELISA)	Functional assay (heparin-induced platelet aggregation or HIPA)[a]	Strategy
Negative	NT	UFH is used for CPB.
Borderline positive (OD ≤ 0.6)	Negative	UFH is used for CPB.
Positive (OD >0.6)	Negative	TPE may be performed daily prior to surgery, whereas UFH can be used for CPB. However, alternative anticoagulant may be needed after surgery.
Positive (OD >0.6)	Positive	Option 1: TPE daily prior to surgery. Then one TPE immediately prior to and one TPE immediately after surgery on the day of surgery may be performed.
		Option 2: Alternative anticoagulant may be used during surgery (first choice: bivalirudin; second choice: lepirudin).

UFH, unfractionated heparin; NT: not tested.
[a]Serotonin Release Assay (SRA) result, if available, can be used in place of HIPA.

complications, especially those with high titers of circulating antibodies shown by HPF4. In this case, plasma exchange (plasmapheresis) can significantly decrease the antibody titer and risk associated with high titer. Plasma exchange is performed with fresh frozen plasma replacement using a 1.0 to 1.5 plasma volume exchange (∼2000−4000 mL based on patient height, weight, gender and hematocrit). The timing of plasmapheresis is dependent on the available time span before surgery. Ideally, therapeutic plasma exchange (TPE) is performed daily (typically for 3 consecutive days before surgery), but if necessary it can be performed just prior to CPB (eg, heart transplant and emergent LVAD). It is important to check HPF4 before and after each TPE. Postoperatively, use of alternative anticoagulants is recommended if the patient requires such medication (first choice, bivalirudin; second choice, lepirudin). It is also advisable to check HPF4 daily with clinical follow-up for up to 4 days after surgery.

- If the patient had positive anti-HPF4 (OD >0.6), positive HIPA, then option 1 is to perform plasmapheresis daily with 1−1.5 plasma volume with FFP daily (typically for 3 consecutive days prior to surgery). It is important to check HPF4 and HIPA before and after each TPE. It is advisable to perform one TPE immediately prior to and one TPE immediately after surgery on the day of surgery. UFH can be used during the CPB, and then the effect of UFH can be reversed using protamine after surgery. Postoperatively, alternative anticoagulants may also be used if the patient needs such therapy (first choice, bivalirudin; second choice, lepirudin). HPF4 and HIPA should be checked on a daily basis with clinical follow-up for 4 days after surgery.

As a second option, alternative anticoagulants may be used during surgery (first choice, bivalirudin; second choice, lepirudin).

The following are important points to remember regarding cardiopulmonary bypass and anticoagulation with UFH for patients with HIT:

- Some patients undergoing cardiovascular surgery may not be exposed to UFH during surgery (eg, patients with right ventricular assist device (RVAD) placement may be kept on ECMO with standard dose of bivalirudin). Communication with the surgery team is important to avoid unnecessary plasmapheresis.
- Patients with a history of HIT waiting for surgery should have HPF4 done twice a week (eg, Monday and Thursday) to monitor the antibody. If HPF4 is positive, then HIPA or SRA testing should be ordered.
- An additional apheresis procedure(s) may be performed if postoperative clinical signs/symptoms are suggestive of heparin-induced thrombocytopenia with thrombosis (HITT) in the setting of acute drop in platelet count and signs of thrombotic complications. In such a scenario, 5% albumin may be used as plasma replacement beyond the 48 h of surgery and the patient does not have bleeding. Plasmapheresis should be performed daily with 1−1.5 plasma volumes daily (typically for 3 consecutive days). It is important to check HPF4 and HIPA before and after each TPE. Rising platelet count is a typical indication of recovery.
- Of course, the decision to proceed with plasma exchange and UFH has to be agreed on by the heart failure team (including surgery, cardiology and anesthesia) for individual cases. Both approaches (preoperative and postoperative) need close coordination by all groups for optimal timing of pheresis.

- At our institution, HPF4 is batched twice daily on weekdays (Monday–Friday) and once on Saturday or Sunday. The cutoff receiving time is 9:00 am and 4:00 pm Monday to Friday and noon on Saturday and Sunday (receiving time). Test results should be available 3 or 4 h after cutoff times. HIPA assay is done during the morning shift. If the testing needs to be done during the weekend, we arrange with the hematology lab manager to have special coagulation technologists available.
- HIT antibody by SRA can be sent to reference laboratories. Some reference laboratories offer 24-h turnaround time.
- If a patient develops marked thrombocytopenia in the postoperative period, days after exposure to UFH in surgery, and currently has no clinical evidence of thrombosis, platelets may be transfused because heparin is not in the patient's circulation to cause thrombotic complications.
- The IgG-specific ELISA was associated with greater specificity (93.5% vs 89.4%) but lower sensitivity (95.8% vs 98.1%) than the polyspecific ELISA [11]. This is attributed to the notion that activation of platelets by HIT antibodies is primarily due to the IgG subclass [12]. The IgG-specific ELISA yields fewer false-positive results than the polyspecific ELISA but at the expense of missing a small proportion of patients with true HIT who are captured by the polyspecific assay [13]. The polyspecific ELISA is used at our institution. Sensitivity and specificity of various HIT tests are listed in Table 7.4. HIT test results can also be combined with other clinical and laboratory results for a more accurate diagnosis of HIT, as seen in Table 7.5 [14].

7.2.9 Anticoagulation With Bivalirudin (Angiomax) During Cardiopulmonary Bypass for Patients With Heparin-Induced Thrombocytopenia

During CPB, there are various key points regarding anticoagulation with bivalirudin (Angiomax) in patients with HIT [15], including the following:
- Bivalirudin dose should be 1 mg/kg (body weight) bolus, followed by 2.5 mg/kg/h. If ACT <400, then another bolus dose of 0.5 mg/kg may be administered followed by 5 mg/kg/h. ACT must be monitored.
- For patients with normal renal function, the half-life of bivalirudin is 25 min. However, bivalirubin clearance is reduced 80% in dialysis-dependent patients, in whom half-life may be increased up to 3.5 h.

Table 7.4 Sensitivity and specificity of HIT tests

Assay category	Mechanism	Examples	Sensitivity (%)	Specificity (%)	Comments
Immunologic	Detects antibodies against PF4/heparin, regardless of their capacity to activate platelets	1. Polyspecific ELISA 2. IgG-specific ELISA	>95	50–89	OD of ELISA result correlates with clinical probability of HIT and odds of a positive functional assay
Functional	Detects antibodies that induce heparin-dependent platelet activation	1. SRA 2. HIPA	90–98	90–95	Not widely available; requires referral to a reference laboratory

HIPA, heparin–induced platelet activation assay; OD, optical density; PF4, platelet factor 4; SRA, serotonin release assay.

Table 7.5 Estimating the pretest probability of HIT: The "four T's"

	Points (0, 1 or 2 for each of four categories; maximum possible score = 8)		
	2	1	0

Pretest probability score: 6−8, high; 4−5, intermediate; 0−3, low

	2	1	0
Thrombocytopenia	>50% platelet fall to nadir ≥ 20	30−50% platelet fall, or nadir 10−19	<30% platelet fall, or nadir <10
Timing[a] of onset of platelet fall (or other sequelae of HIT)	Days 5−10, or ≤ Day 1 with recent heparin (past 30 days)	>Day 10 or timing unclear; or <Day 1 with recent heparin (past 31−100 days)	<Day 4 (no recent heparin)
Thrombosis or other sequelae	Proven new thrombosis; skin necrosis; or acute systemic reaction after intravenous UFH bolus	Progressive or recurrent thrombosis; erythematous skin lesions; suspected thrombosis (not proven)	None
Other cause(s) of platelet fall	None evident	Possible	

[a]First day of immunizing heparin exposure is considered Day 0.

- During CPB, the ACT value should be kept in the range 400−500 s, TEG R 20−25, INR value approximately 8 and PTT approximately 180 s.
- Two hours after the end of bivalirudin infusion, expected ACT should be less than 270 s, PTT should be less than 100 s and INR should be less than 2.4.
- Clearance of bivalirudin can be confirmed with a normal thrombin time (<21 s).
- If the patient is bleeding actively due to bivalirubin, FFP, cryoprecipitate and rFVIIa (last resort) can be attempted together with renal dialysis [16]. Platelets can also be used if there are qualitative or quantitative platelet defects.

It is important to note that the use of bivalirubin in CPB has been known to be associated with severe bleeding complications in some cases.

7.2.10 Anticoagulation Following Bioprosthetic Valve Implantation

A variety of combinations (antiplatelet only, Coumadin only or both) have been used for different types and combinations of valve repairs (atrial valve and mitral valve) with thromboembolic risk factors in consideration. Thromboembolic risk factors include atrial fibrillation, prior thromboembolism, left ventricular dysfunction (ejection fraction $<30\%$) and hypercoagulable state. Refer to Tables 7.6 and 7.7 for details [17, 18].

7.3 DIFFICULT PROBLEMS IN COAGULOPATHY MANAGEMENT

In this section, difficult problems associated with coagulopathy in various clinical scenarios involving critically ill patients are discussed (Table 7.8).

7.3.1 Anticoagulant for Patients With Thrombotic Complications and Hypocoagulopathy Without Active Bleeding

Occasionally, difficulty in anticoagulant management may be encountered for patients who need to be started on anticoagulant for thrombotic complications (eg, showing loss of arterial pulse) and have existing

Table 7.6 Discharge anticoagulants and antiplatelets by valve type for patients without risk factors

	Overall (%)	AVR (%)	MVR (%)	AVR + MVR (%)
Antiplatelet only	74.8	77.7	37.5	66.7
Warfarin only	3.3	1.8	25.0	0.0
Both	13.0	12.5	25.0	0.0
None	8.9	8.0	12.5	33.3

AVR, aortic valve replacement; MVR, mitral valve replacement.

Table 7.7 Discharge anticoagulants and antiplatelets by valve type for patients with risk factors

	Overall (%)	AVR (%)	MVR (%)	AVR + MVR (%)
Antiplatelet only	48.7	51.5	35.0	0.0
Warfarin only	15.7	13.9	35.0	0.0
Both	33.5	32.4	30.0	100.0
None	2.0	2.3	0.0	0.0

AVR, aortic valve replacement; MVR, mitral valve replacement.

Table 7.8 A typical modified bivalirudin protocol (therapeutic PTT 50—65 s rather than 50—80 s)

PTT range (s)	Action
<50	Increase bivalirudin rate by 20%.
50—65	No action (therapeutic range).
65—80	Reduce bivalirudin rate by 25%.
81—105	Reduce bivalirudin rate by 50%.
>105	Hold bivalirudin for 1 h and reduce rate by 50%.
>200	Hold bivalirudin for 2 h and then check stat PTT. Resume dosing per protocol with repeat PTT.

hypocoagulopathy (prolonged PT/aPTT, etc.). If patients do not have active bleeding, therapy with anticoagulant (bivalirudin) should be started with low dose 0.005 mg/kg/h. UFH should be avoided due to possible HIT causing thrombosis. Interestingly, infusion rate in this range does not prolong PTT to any noticeable degree. If INR is significantly prolonged (INR >2.5), as typically seen in liver failure, patients may be transfused with FFP to decrease INR (down to ~2.0) to prevent bleeding before slowly increasing the bivalirudin dose to a therapeutic PTT. Transfusion with FFP and bivalirudin dose adjustment should be coordinated.

7.3.2 HLA Antibody Workup for Refractory Thrombocytopenia Due to HLA Alloantibodies

For patients with refractory thrombocytopenia with proven lack of response to platelet transfusion, workup for HLA alloantibodies may be initiated with consultation with the blood bank attending physician. With consensus by the blood bank attending physician, these steps may be followed to facilitate a speedy workup:

- One blood specimen collected in a red-top tube should be sent to the laboratory for testing of HLA antibody panel. Specific clinical history should be included in the request form (refractory thrombocytopenia, not responding to platelet transfusion). Optional HLA typing can also be ordered with specimen collected in a yellow-top tube.
- For refractory thrombocytopenia, the most important data are antibodies against HLA-A, HLA-B (class 1). Titer [mean fluorescent intensity (MFI)] against these antibodies indicates the strength of the antibodies. A high percent reactive antibody indicates increasing difficulty in getting compatible platelets. The report shows an antibody table with MFI (titer) against HLA-A and HLA-B loci. Details are

given regarding unacceptable donor and acceptable donor for platelet transfusion. However, even for those with MFI >500, some immune suppression measures are suggested with the platelet transfusion, such as high-dose intravenous steroid, immunoglobulins and rituximab.

- The HLA report can be sent to the American Association of Blood Banks (AABB) with a request for number of platelet units (eg, one dose each day). AABB is to send compatible platelets to the regional blood center to be distributed to the hospital.

7.3.3 Management of Transfusion-Dependent Patients on ECMO

Critically ill patients may become transfusion-dependent. A typical case is a patient who presents with cardiogenic shock, intubated, status post ECMO, shocked liver, acute renal failure, on continuous renal replacement therapy and coagulopathy with DIC requiring multiple blood transfusions. A similar scenario is a patient with LVAD and RVAD who is experiencing shocked liver and renal failure. Hypocoagulopathy should be corrected with the following transfusion thresholds:

- If platelet count is less than $50,000/\mu L$, then platelet transfusion is indicated.
- If hemoglobin is less than 8 g/dL, then RBC transfusion is indicated.
- If INR >2.5, then transfusion with FFP may be indicated.
- Fibrinogen <150, transfusion with FFP or cryoprecipitate may be indicated.

If the patient is not bleeding, heparin should be used for ECMO with the therapeutic goal of PTT approximately 55−75 s. If INR is prolonged, low-dose heparin may be used. ASA does not need to be discontinued. Other antiplatelet medications (typically ADP-P2Y12 inhibitors) need to be discontinued with patients on ECMO.

If the patient has thrombosis and no acute bleeding, bivalirudin is a better choice (dose for regular acute coronary syndrome—ACS protocol). Dose is adjusted using bivalirudin protocol with typical therapeutic PTT of 50−80 s. If the patient has prolonged baseline PT/PTT, a low dose of bivalirudin (eg, 0.02 mg/kg/h) may be more appropriate. Dose can be decreased much lower as needed (eg, 0.005 mg/kg/h) with specific order for no titration as needed to prevent excessively prolonged PTT, especially for patients with renal insufficiency. Patients need bivalirudin to inhibit thrombin in ongoing thrombosis. If INR is significantly prolonged (INR >2.5), typically seen in liver failure, transfusion with FFP in order to decrease INR (down to ~2.0) may be advisable before slowly increasing the bivalirudin dose. FFP transfusion and bivalirudin dose adjustment need to be coordinated together. Platelet count needs to be kept over 50,000/µL. FFP or cryoprecipitate transfusion may be needed if fibrinogen is less than 150.

If the patient is bleeding (with or without thrombosis), no anticoagulant should be used, especially if INR is already elevated. FFP is required if INR is significantly prolonged (>2.0). The targeted INR is ≤2.0. Antiplatelet medications (ASA) should be discontinued. Platelet count needs to be maintained over 60,000/µL. Once bleeding is resolved in the patient, bivalirudin (eg, 0.005 mg/kg/h, with no titration) can be started at low dose so that PTT is not significantly prolonged. Bivalirudin can later be scaled up to regular protocol with titration to maintain PTT in therapeutic range when bleeding is completely resolved. FFP transfusion and bivalirudin dose adjustment need to be coordinated. CBC should be ordered along with platelet count and DIC screen every 2—4 h.

For significant thrombocytopenia (platelet count <20,000/µL), even without bleeding, antiplatelet medications need to be on hold until platelet count recovers. Anticoagulation in patients with marked thrombocytopenia (platelet count <20,000/µL) may be temporarily ceased in these patients to prevent bleeding.

For patients who are critically ill with ongoing coagulopathy, a standing order for laboratory testing and transfusion would simplify the process. For example, for a nonbleeding patient, CBC without differential, platelet count, and DIC screen should be ordered every 4 h. If the hemoglobin is less than 8.0 g/dL, transfusion with RBC is helpful. If INR is over 2.5, then transfusion with RBC may be indicated. If platelet count falls below 50,000/µL, the pathologist on the hemotherapy service should be consulted.

For a bleeding patient, CBC without differential, platelet count and DIC screen should be ordered every 2 h. If hemoglobin is less than 8.0 mg/dL, then transfusion with RBC may be appropriate. If INR is over 2.0, then FFP should be given. If platelet count is less than 60,000/µL, the pathologist on hemotherapy service should be contacted.

Typically, RBCs are used most often, followed by platelets and FFP. RBCs can be transfused through the ECMO line. Other blood components have to be transfused to patients through another access lines. If PTT is not responding to heparin, ATIII level should be checked. If ATIII is very low, FFP or ATIII concentrate may be administered. However, if PTT level responds to heparin, there is no additional need to check ATIII level.

7.3.4 Thrombophilia Workup

Patients with significant clinical findings of thrombophilia need further investigation with the following laboratory tests: FV Leiden, FII mutation, lupus anticoagulant, anticardiolipin antibodies, protein C, protein S, antithrombin, homocysteine and HIT (with supporting clinical and laboratory findings). If any of these tests is positive, clinical hematology consult should be suggested to the heart failure team. Note that the levels of protein C and protein S would be decreased with Coumadin. UFH and direct thrombin inhibitors may yield a false-positive result for lupus anticoagulant.

7.3.5 Modifying Anticoagulant Protocol for Patients With Risk of Both Bleeding and Thrombosis

Some patients may be at risk for both bleeding and thrombosis, and a modified anticoagulant protocol may be needed for these patients. For example, a modified bivalirudin protocol may be needed due to a new pulmonary embolic event in a patient with bleeding risk (therapeutic PTT 50–65 sec rather than 50–80 sec). Consider the case of a patient currently on bivalirudin 0.01 mg/kg/h with PTT 47 s. The bivalirudin infusion rate may be changed to 0.015 mg/kg/h to maintain PTT above 50 s, and then a new modified protocol (typically created with assistance from the cardiovascular pharmacy team) may be used for further management.

A starting dose for bivalirudin is typically 0.05 mg/kg/h. PTT values should be assessed every 4 h until they are within therapeutic range. PTT values should also be monitored after any rate change. When two consecutive PTT values are within therapeutic range, further PTT values may be determined every 12 h while the patient is on bivalirudin therapy.

Similarly for UFH infusion, if the patient has not been started on UFH, initial infusion rate should be selected and then the infusion rate may be increased slowly (by 200 units/h) to obtain PTT values in the desired therapeutic range. After that, a modified UFH protocol may be initiated.

7.3.6 Management of HITT in Bleeding Patients

Management of HITT may be problematic in bleeding patients (eg, intracranial bleeding). It has been demonstrated that plasma exchange is a useful alternative to anticoagulant (bivalirudin) in such clinical presentation. Plasma exchange (1 plasma volume with FFP) can remove the PF4/heparin complexes, thus preventing ongoing thrombosis and allowing platelet recovery. This deters any further bleeding complications, allowing for implementation of appropriate anticoagulation. The effects of HIT antibodies and the risk for prothrombotic complications can be significantly reduced after two procedures, with no further evidence of HIT after four procedures.

7.3.7 Monitoring Anticoagulant in Patients With Lupus Anticoagulant and Severe Liver Disease

In such patients, PTT is prolonged due to the combined effect of underlying liver dysfunction, lupus anticoagulant and anticoagulant such as bivalirudin. Lupus anticoagulant is likely to cause a prolonged baseline PTT. However, alternate testing (dilute thrombin time) is not useful with underlying liver disease. PTT is the best option to monitor bivalirudin dosing, and PTT should be maintained at the high end of the therapeutic range to compensate for the prolongation of baseline PTT due to lupus anticoagulant. A modified anticoagulant protocol may be needed. For example, a modified bivalirudin protocol with therapeutic PTT 60—100 s rather than 50—80 s may be targeted.

7.3.8 Modifying Anticoagulant Protocol for Patients With Severe Thrombophilia Despite Therapeutic PTT for UFH

For patients on UFH with severe thrombosis despite having therapeutic PTT, a modified anticoagulant protocol may be needed. For example, a modified UFH protocol due to repeated thrombotic events would use a therapeutic range PTT of 70—100 s rather than 55—75 s. Modification of UFH protocol is similar to that described in the previous section.

7.3.9 Transfusion Support for DIC

Consumption coagulopathy in DIC presents a challenge for management [19]. In patients with bleeding, or in the immediate post-op period with abnormal coagulation parameters, transfusion is needed to control hemostasis. However, in patients without bleeding or not in the immediate post-op period, prophylactic transfusion is not indicated except for very critical coagulation parameters that predispose patients to spontaneous bleeding—that is, platelet count <15,000—20,000/μL, INR >3.0—3.5 and fibrinogen <100. Antifibrinolytic agents are needed for primary fibrinolysis but are contraindicated in secondary fibrinolysis (early phase

Table 7.9 DIC score by ISTH

Risk assessment: Does the patient have an underlying disorder known to *be* associated with overt DIC?

If yes: proceed

If no: do not use this algorithm

Order global coagulation tests (PT, platelet count, fibrinogen, fibrin-related marker)

Score the test results
- Platelet count ($>100 \times 10^9$/L = 0, $<100 \times 10^9$/L = 1, $<50 \times 10^9$/L = 2)
- Elevated fibrin marker (eg, D-dimer, fibrin degradation products) (no increase = 0, moderate increase = 2, strong increase = 3)
- Prolonged PT (<3 s = 0, >3 but <6s = 1, >6s = 2)
- Fibrinogen level (>1 g/L = 0, <1g/L = 1)

Calculate score

≥ 5 compatible with overt DIC: repeat score daily

<5 suggestive for nonovert DIC: repeat next 1–2 days

of DIC). Diagnosis of DIC can be more accurately obtained with the scoring system by the International Society for Thrombosis and Haemostasis (ISTH) shown in Table 7.9 [20].

7.4 CONCLUSIONS

This chapter covered a wide range of special topics in the management of coagulopathy that were not discussed in previous chapters. Most of these topics involve critically ill patients who require mechanical assist devices such as VAD and ECMO. Assessment of coagulopathy and clinical management may be difficult due to many confounding factors. Practical measures in managing such patients were outlined with references cited for readers who need to acquire more details on the specific topics.

REFERENCES

[1] Rossi M, Serraino GF, Jiritano F, Renzulli A. What is the optimal anticoagulation in patients with a left ventricular assist device? Interact Cardiovasc Thorac Surg 2012;15:733–40.

[2] Pucher PH, Cummings IG, Shipolini AR, McCormack DJ. Is heparin needed for patients with an intra-aortic balloon pump? Interact Cardiovasc Thorac Surg 2012;15:136–9.

[3] Ageno W, Gallus AS, Wittkowsky A, Mark Crowther M, et al. Oral anticoagulant therapy antithrombotic therapy and prevention of thrombosis, 9th edition: American College of Chest Physicians evidence-based clinical practice guidelines. Chest 2012;141:e44S–88S.

[4] Koster A, Weng Y, Battcher W, Gromann T, et al. Successful use of bivalirudin as anticoagulant for ECMO in a patient with acute HIT. Ann Thorac Surg 2007;83:1865—7.

[5] Ensor CR, Cahoon WD, Crouch MA, Katlaps GJ, et al. Antithrombotic therapy for the CardioWest temporary total artificial heart. Tex Heart Inst J 2010;37:149—58.

[6] The A.S.P.E.N. Nutrition Support Patient Education Manual. American Society for Parenteral & Enteral Nutrition. http://www.nutritioncare.org.

[7] Hadem J, Hafer C, Schneider AS, Beutel G, et al. Therapeutic plasma exchange as rescue therapy in severe sepsis and septic shock: retrospective observational single-centre study of 23 patients. BMC Anesthesiol 2014;14:24.

[8] Costanzo MR, Dipchand A, Straling R, Anderson A, et al. The International Society of Heart and Lung Transplantation guidelines for the care of heart transplant recipient. J Heart Lung Transplant 2010;29:914—56.

[9] Selleng S, Haneya A, Hirt S, Selleng K, et al. Management of anticoagulation in patients with subacute heparin-induced thrombocytopenia scheduled for heart transplantation. Blood 2008;112:4024—7.

[10] Welsby IJ, Um J, Milano CA, Ortel TL, et al. Plasmapheresis and heparin reexposure as a management strategy for cardiac surgical patients with heparin-induced thrombocytopenia. Anesth Analg 2010;110:30—5.

[11] Cuker A, Ortel TL. ASH evidence-based guidelines: is the IgG specific anti-PF4/heparin ELISA superior to the polyspecific ELISA in the laboratory diagnosis of HIT? ASH Education Book. Hematology 2009;251—2. American Society of Hematology.

[12] Vun CM, Evans S, Chesterman CN. Anti-PF4-heparin immunoglobulin G is the major class of heparin-induced thrombocytopenia antibody: findings of an enzyme-linked immunofiltration assay using membrane-bound hPF4-heparin. Br J Haematol 2001;112:69—75.

[13] Cuker A., Crowther M.A. 2013 Clinical Practice Guideline on the Evaluation and Management of Adults with Suspected Heparin-Induced Thrombocytopenia (HIT). American Society of Hematology.

[14] Warkentin TE. Heparin-induced thrombocytopenia: diagnosis and management. Circulation 2004;110:e454—8.

[15] Vasquez JC, Vichiendilokkul A, Mahmood S, Baciewicz Jr. A. Anticoagulation with bivalirudin during cardiopulmonary bypass in cardiac surgery. Ann Thorac Surg 2002;74:2177—9.

[16] Stratmann G, de Silva AM, Tseng EE, Hamblenton J, et al. Reversal of direct thrombin inhibition after cardiopulmonary bypass in a patient with heparin-induced thrombocytopenia. Anesth Analg 2004;98:1635—9.

[17] Brennan JW, Alexander KP, Wallace A, Audra B, et al. Patterns of anticoagulation following bioprosthetic valve implantation: observations from ANSWER. J Heart Valve Dis 2012;21:78—87.

[18] Bartholomew JR. Transition to an oral anticoagulant in patients with heparin-induced thrombocytopenia. Chest 2005;127(2 Suppl.):27S—34S.

[19] Gullo A, et al. Chapter 7: Disseminated intravascular coagulation. In: Giorgio Berlot, editor. Hemocoagulative problems in the critically Ill patients. Springer; 2012. p. 93—107.

[20] Levi M, Toh CH, Thachil J, Watson HS, et al. Guidelines for the diagnosis and management of disseminated intravascular coagulation. Br J Haematol 2009;145: 24—33.

CHAPTER 8

Decision Support Software for Coagulopathy

Contents

8.1 INTRODUCTION

This chapter highlights decision support software that may be used as a tool for the management of bleeding patients. Two program applications designed and implemented by our group are described:
- Excel file for transfusion dosage based on laboratory results.
- An artificial neural network for predicting blood use.

8.2 MOBILE COMPUTING PLATFORM WITH DECISION SUPPORT MODULES FOR HEMOTHERAPY

This software is based on a Microsoft Excel (Microsoft, Redmond, WA) spreadsheet template [1]. The template is used intraoperatively for consultation when a patient is undergoing cardiopulmonary bypass (CPB). The goal is to identify any potential hemostatic defect in a patient so that the correct blood components may be used for transfusion if needed. In addition, dosage of individual blood products may also be selected depending on the severity of bleeding and abnormal parameters of the coagulation test results. In the template, the pathologist starts with data entry fields (DEFs) to enter pertinent information about the patient, including general demographic identifiers, medications, preoperative laboratory results, and coagulation risk assessment results (Fig. 8.1).

A. Nguyen, A. Dasgupta, A. Wahed:
Management of Hemostasis and Coagulopathies for Surgical and Critically Ill Patients.
DOI: http://dx.doi.org/10.1016/B978-0-12-803531-3.00008-2

Drop-down menus are available in certain fields, such as names of the attending physicians, to increase the speed of data entry into the spreadsheet. In the next section, there are DEFs in which the pathologist can record data for the aforementioned set of tests done on the patient (Fig. 8.2). The test result fields are automatically highlighted if the values

HEMOTHERAPY DATA SHEET								
							Ver 8/11/2013	
Consult Type:	LVAD		Pathologist:					
Timing:	operative		Path resident:					
Location:	OR #1		Cardiologist:					
Date:	xx/xx/xxxx		Surgeon:					
Patient: xxxx	MRN: xxxx		Anesthesiologist:					
Medications:	Heparin for Tandem Heart							
Med stopped on:	4 hrs prior to surgery							
LFT results:	Normal		Preop Lab values					
			Hgb:	8.4	Plt Agg:			
Renal function results:	Cr 0.8, BUN 9		Plt:	167	Conc	High	Low	
Chronic renal failure?	no		PT/PTT:	14.8 / 55	AA:			
			Fib:		ADP:			
Coagulation Risk:	intermediate		ATIII:		Coll:			
					EPI:			
					Rist:			

Figure 8.1 Patient's general information (data for a fictitious patient).

							Flag:	HIGH
			[optional]					LOW
Event		#1	#2	#3	#4	#5		
Phase		Baseline	before off-pump	Off-pump	Closing	Closing		
Time		9:30	11:45	13:40	15:20	16:50		
R	[5-10] min	xxxx	xxxx					
h-R	[5-10] min		5.8					
h-Alpha	[53-72] degree		67					
h-MA	[50-70] mm		64.4					
h-EPL	[0-15] %		0					
h-Ly30	[0-8] %		0					
h-CI	[-3.0-3.0]		0.7					
Hgb	[12-18] g/dL	7.9	8.1	6.7	8.5	10.1		
Plt	[133-450]x10³ /µL	167	91	55	153	127		
PT	[12-14.7] sec	14.7	22.9	28.4	20.6	21.1		
PTT	[22.9-35.8] sec	33.1	xxxx	210	48.7	43.1		
Fib	[230-510] mg/dL	513	261	186	342	273		
TT	[15.0-21.2] sec	19	xxxx	110	18.9	17.1		
D-Dimer	[0-0.66] µg/mL FEU	0.81		0.81	0.95	0.86		
ATIII	[77-140] %	82	xxxx	xxxx	xxxx	xxxx		
VerifyNow-ASA (#1 only)	[target:>550 ARU]	597	597	597	597	597		
VerifyNow-P2Y12 (#1 only)	[target:>210 PRU]	374	374	374	374	374		

Figure 8.2 Patient's coagulation test results (data for a fictitious patient).

are abnormal (green for low values and red for high values). Here, the values entered include reaction time (R) [from the thromboelastography (TEG)]; heparinase reaction time (h-R), heparinase angle (h-A), heparinase maximum amplitude (h-MA), heparinase estimated percent lysis (h-EPL), heparinase lysis at 30 min (h-Ly30), and heparinase coagulation index (h-CI) (from the h-TEG); hemoglobin (Hgb) and platelet count (from the complete blood count); prothrombin time (PT)/partial thromboplastin time (PTT), fibrinogen, thrombin time, and D-dimer (from the disseminated intravascular coagulation screen); antithrombin (AT); and VFN-A (VerifyNow aspirin test result) and VFN-P (VerifyNow Plavix test result). These tests are sequentially performed and entered in the Excel spreadsheet during different phases of CPB, including baseline, hemoconcentration/rewarming (just prior to termination of CPB), post-CPB, and the potential postoperative bleeding phase. Data validation for input entry is also provided. If the input is not in an acceptable range or it is in an incorrect format, the data entry is stopped with a warning message. As the user enters laboratory results into the template, the values are automatically routed through the decision support module (DSM) in the Excel file. The DSM comprises 45 algorithms that are translated to Excel functions. The physicians in hemotherapy service at our institute, consisting of hematopathologists and blood bankers, developed the algorithms for the DSM by reviewing the literature for algorithms used during CPB surgery; hence, the algorithms are based on standard transfusion practice. Our group prepared a comprehensive hemotherapy guide, which includes all the algorithms, and the current version of this guide can be reviewed on our web site at http://hemepathreview.com.

The algorithms proposed in our guide are continuously revised based on experience learned on our service. The guide and the algorithms within them are also annually reviewed and approved by the hospital's Advanced Heart Failure Clinical Committee. The algorithms consist of 36 level 1 algorithms (Table 8.1) and 9 level 2 algorithms (Table 8.2). Level 1 algorithms provide transfusion needs based on specific sets of laboratory tests. Level 2 algorithms combine results from level 1 algorithms for final transfusion suggestions. These recommendations appear under "Suggestion Summary" (Fig. 8.3) and consist of the following fields: fresh frozen plasma (FFP; single units), cryoprecipitate (doses), platelets (doses or apheresis units), red blood cells (RBCs; units), tranexamic acid (grams), protamine (milligrams), alert for use of factor VIIa (FVIIa), alert for use of AT concentrate, and alert for possible protamine overdose. The 36 level 1

Table 8.1 Level 1 algorithms (total 36 algorithms)

	Algorithm
1. RBCs for anemia	If Hgb < 10 → RBC transfusion to increase Hgb to 10.0 (1 RBC increased Hgb by 1.0)
2. FFPs for low clotting factors	If h-R: 10−15 → 2 units of FFP
3. FFPs for low clotting factors	If h-R: 15−20 → 4 units of FFP
4. FFPs for low clotting factors	If h-R > 20 → 6 units of FFP
5. FFPs for low fibrinogen	If Fib: 150−200 → 2 units of FFP
6. Cryo for very low fibrinogen	If Fib < 150 → 1 dose of cryoprecipitate
7. FFPs for low fibrinogen	If h-alpha: 20−45, h-MA > 50 → 2 units of FFP
8. Cryo for very low fibrinogen	If h-alpha < 20, h-MA > 50 → 1 dose of cryoprecipitate
9. FFPs for low clotting factors	If PT: 20−25 → 2 units of FFP
10. FFPs for low clotting factors	If PT > 25 → 4 units of FFP
11. FFPs for low clotting factors	If PTT: 45−50 → 2 units of FFP (however, if due to heparin effect in algorithm 13, then disregard value), not applicable for on-pump time
12. FFPs for low clotting factors	If PTT > 50 → 4 units of FFP (however, if due to heparin effect in algorithm 13, then disregard value), not applicable to on-pump time
13. No FFP due to heparin effect	If PTT > 45, TT > 25 → due to heparin effect
14. FFPs for low ATIII	If ATIII: 35−50 → 2 units FFP, applicable to baseline only
15. ATIII conc for very low ATIII	If ATIII < 35 → ATIII concentrate, applicable to baseline only
16. Apheresis Plts for low platelets	If Plt: 50−100 → 2 apheresis units of platelets
17. Apheresis Plts for low platelets	If Plt < 50 → 3 apheresis units of platelets
18. Apheresis Plts for low platelets	If h-MA: 35−45, h-EPL < 15, h-Ly30 <8 → 2 apheresis units of platelets
19. Apheresis Plts for low platelets	If h-MA < 35, h-EPL < 15, h-Ly30 < 8 → 3 apheresis units of platelets
20. Apheresis Plts for ADP inhibition	If VFN-P: 130−210 → 2 apheresis units of platelets
21. Apheresis Plts for ADP inhibition	If VFN-P < 130 → 3 apheresis units of platelets
22. Apheresis Plts for AA inhibition	If VFN-A: 350−550 → 2 apheresis units of platelets

(Continued)

Table 8.1 (Continued)

	Algorithm
23. Apheresis Plts for AA inhibition	If VFN-A $< 350 \rightarrow 3$ apheresis units of platelets
24. Tranexamic acid for primary fibrinolysis	If h–EPL > 15, h–MA $< 50 \rightarrow 1$ g of tranexamic acid
25. Tranexamic acid for primary fibrinolysis	If h–EPL > 15, h–CI $< 1 \rightarrow 1$ g of tranexamic acid
26. Tranexamic acid for primary fibrinolysis	If h–Ly30 > 8, h–MA $< 50 \rightarrow 1$ g of tranexamic acid
27. Tranexamic acid for primary fibrinolysis	If h–Ly30 > 8, h–CI $< 1 \rightarrow 1$ g of tranexamic acid
28. Tranexamic acid for primary fibrinolysis	If D–dimer > 10, Fib $< 150 \rightarrow 1$ g of tranexamic acid
29. Protamine for excess heparin	If PTT > 45, TT $> 25 \rightarrow 50$ mg of protamine (only off pump and after)
30. Secondary fibrinolysis, no FVIIa	If h–EPL > 15, h–MA $> 70 \rightarrow$ no FVIIa
31. Secondary fibrinolysis, no FVIIa	If h–EPL > 15, h–CI $> 3 \rightarrow$ no FVIIa
32. Secondary fibrinolysis, no FVIIa	If h–Ly30 > 8, h–MA $> 70 \rightarrow$ no FVIIa
33. Secondary fibrinolysis, no FVIIa	If h–Ly30 > 8, h–CI $> 3 \rightarrow$ no FVIIa
34. Possible protamine overdose	If h–TEG R > 20, h–alpha < 20, h–MA < 35 (only off pump and after, no bleeding seen) $\rightarrow$ yes to possible protamine overdose
35. Cryo for chronic renal failure	Pos Hx of CRF (uremia)
36. Only baseline VFNs needed	If baseline VFN exists, carry over to subsequent phases

algorithms can be divided into 10 different product/recommendation categories: (1) RBC transfusion for low hemoglobin; (2) FFP transfusion for low clotting factors, low AT, or low fibrinogen; (3) cryoprecipitate transfusion for very low fibrinogen or uremia; (4) AT concentrate transfusion for very low AT; (5) no FFP transfusion for prolonged PTT secondary to heparin; (6) platelet transfusion for low platelet count or medication effect; (7) tranexamic acid for primary fibrinolysis; (8) protamine for excess heparin after neutralization; (9) FVIIa contraindicated in secondary fibrinolysis; and (10) alert for possible protamine overdose.

Table 8.2 Level 2 algorithms (9)

	Summary of components	Rules to combine algorithm steps
I	FFPs (single units)	Maximum value from algorithms 2−5, 7, 9−12, 14
II	Cryo (dose)	Maximum value from algorithms 6, 8, 35
III	Plts (apheresis units)	Maximum value from algorithms 16−23
IV	RBCs (units)	RBCs to increase Hgb to 10: A RBCs to correct Hgb for FFP: $B = 0.5 \times FFPs$ RBCs to correct Hgb for Plts: $C = Plts$ If Hgb−(A + B) > 10, no RBCs needed Otherwise, RBCs = A + B + C
V	EACA (grams)	Maximum value from algorithms 24−28
VI	Tranexamic acid (grams)	Value from algorithm 29
VII	ALERT for use of FVIIa (contraindicated)	Any positive value from algorithms 30−33
VIII	ALERT for AT conc	Value from algorithm 15
IX	ALERT for protamine overdose	Value from algorithm 34

		A	B	C	D	E	F	G	H	I
44	Suggestion Summary: I FFPs (single units)		0	2	4	2	2			
45	II Cryo (dose)		0	0	0	0	0			
46	III Plts (apheresis units)		0	2	2	0	0			
47	IV RBCs (units):		2	5	7	3	1			
48	V Tranexamic acid (gm):		0	0	0	0	0			
49	VI Protamine (mg):		xxxx	xxxx	50	0	0			
50	VII ALERT for use of FVIIa:		---	---	---	---	---			
51	VIII ALERT for AT conc:		---							
52	IX ALERT for Protamine Overdose:			---	---	---				

Figure 8.3 Summary of suggestions for management (data for a fictitious patient).

In general, fibrinogen concentrate has not been used by our service because we find cryoprecipitate adequate to restore fibrinogen level. Furthermore, no DSM has been developed for the new anticoagulants dabigatran, rivaroxaban, and apixaban, even though we find these anticoagulants problematic for emergency cardiac surgery. Instead, we have guidelines for monitoring these medications and alleviating bleeding in our online guide. For illustration, let us examine a clinical scenario at the time of off-pump (discontinuation of CPB) with the following laboratory results: Hgb, 6.7 g/dL; platelets, 55,000/μL; PT, 28.4 s; PTT, 210 s; fibrinogen, 186 mg/dL; and thrombin time, 110 s. When the user enters data into the DEF, the system routes those values to the DSM and considers the following level 1 algorithms:

- Algorithm 1: Recommend transfusion with three units of RBCs to increase Hgb from 6.7 g/dL to above 10 g/dL.
- Algorithm 5: Recommend transfusion with two units of FFPs for low fibrinogen at 186 mg/dL.
- Algorithm 10: Recommend transfusion with four units of FFPs for prolonged PT at 28.4 s.
- Algorithm 13: No FFP needed for prolonged PTT at 210 s (this is due to the heparin effect with thrombin time at 110 s).
- Algorithm 16: Two doses of platelets for platelet count at 55,000/μL.
- Algorithm 29: Administration of protamine (50 mg) for excess heparin (prolonged PTT and thrombin time).

Level 2 algorithms are then activated to combine results from level 1 algorithms for the actual number of blood components needed:

- Number of units of FFPs: Value from algorithm 5 is two units and from algorithm 10 is four unites; therefore, a maximum of four units may be transfused.
- Number of units of platelets: Two units from algorithm 16; therefore, a maximum of two units of platelets may be transfused.
- Number of units of RBCs
  - to correct for FFP transfusion $= 0.5 \times$ FFPs $= 0.5 \times 4 = 2$ units,
  - to correct for platelet transfusion $=$ platelets $= 2$ units,
  - to correct for low Hgb: From algorithm 1 $= 3$ units.

    Therefore, total units of RBCs $= 2 + 2 + 3 = 7$ units.
- Protamine: From algorithm 29 $= 50$ mg.

The final results from level 2 algorithms are displayed in the "Suggestion Summary" (Fig. 8.3). Note that fields such as Cryoprecipitate and Tranexamic Acid are left blank because their associated algorithms (both level 1 and level 2) have not been activated with the given laboratory results. As the laboratory data of the patient are entered, the clinician can review the recommendations produced by the DSM and combine them with his or her clinical judgment to make a final decision for transfusion. It cannot be overemphasized that clinical judgment is critical for a final transfusion decision based on the clinical situation. In particular, the number of blood components should be based on the degree of microvascular bleeding and the patient's body weight. The dosage recommendation from our DSM can be used as a starting point for further adjustment. The suggested transfusion is most applicable for the off-pump phase, when most patients will exhibit some degree of microvascular bleeding. For suggested transfusion shown in other phases (baseline and hemoconcentration), transfusion typically is not needed. The suggestions for these phases just need to be kept in mind for later management.

8.3 POSTOPERATIVE BLEEDING RISK STRATIFICATION IN CARDIAC PULMONARY BYPASS PATIENTS USING ARTIFICIAL NEURAL NETWORK

Artificial neural network (ANN) is software designed with parallel distributed components or adaptive systems [2]. They are composed of a series of interconnected processing elements that operate in parallel. Neural networks lack centralized control in the classical sense because all the interconnected processing elements change or "adapt" simultaneously with the flow of information and adaptive rules. Neural networks can model highly nonlinear systems in which the relationship among the variables is unknown or very complex. The network is trained using a set of input—output pairs. For each example in the training set, the network receives an input and produces an actual output. After each trial, the network compares the actual with the desired output and corrects any difference by slightly adjusting all the weights in the network until the output produced is similar enough to the desired output or the network cannot improve its performance any further.

The prediction of bleeding risk in CPB patients has a vital role in their postoperative management. Thus, an ANN to analyze intraoperative laboratory data to predict postoperative bleeding was designed and implemented by our group. The JustNN software (Neural Planner Software, Cheshire, United Kingdom) was used to design our ANN. This ANN was trained using 15 intraoperative laboratory parameters paired with one output category, risk of bleeding, defined as units of blood components transfused in 48 h from the start of surgery. The ANN was trained with the first 39 CPB cases. The set of most important input parameters for this ANN was also determined, and the ANN was validated using the next 13 cases. The set of most input parameters includes five components: prothrombin time, platelet count, thromboelastograph reaction time, D-dimer, and thromboelastograph coagulation index (Fig. 8.4). The validation results show nine cases (69.2%) with exact match and three cases (23.1%) with a one-grading difference and one case (7.7%) with a two-grading difference between actual blood usages and predicted blood usage.

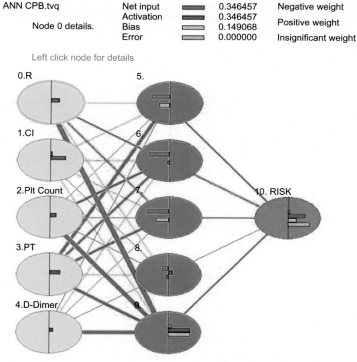

ANN CPB.tvq

Node 0 details.

Left click node for details

Net input	▬	0.346457	Negative weight ▬▬▬▬
Activation	▬	0.346457	Positive weight ▬▬▬▬
Bias	▭	0.149068	
Error	▬	0.000000	Insignificant weight -------

0.R
1.CI
2.Plt Count
3.PT
4.D-Dimer

5.
6.
7.
8.
9.

10. RISK

Figure 8.4 Visual schematics of our neural network. The first column of spheres represents the input parameters, the second column of spheres represents the hidden layer, and the last column of spheres represents the output parameter. Each sphere represents a node, and the lines between the spheres represent the weights of the connections between the nodes.

To the best of our knowledge, this ANN is the first one developed for postoperative bleeding risk stratification of CPB patients. With promising results, we have started using this ANN to risk-stratify our CPB patients, and it has assisted us in predicting postoperative bleeding risk.

Instruction for using neural network to predict risk of perioperative and postoperative bleeding:

Download and install software JustNN at http://www.justnn.com.

Download the trained Neural Network data file from http://hemepa-threview.com/CBH/LVAD + OHT-Cases-5-Inputs.tvq.

Open program JustNN, open data file, click on "Query" in task bar, select "Add query," double-click on the first input field, enter result, repeat for the remaining four inputs (see Fig. 8.5).

	R	CI	Plt Count	PT	D-Dimer	RISK	
Query	?	?	?	?	?	?	

Figure 8.5 Grid view of data input and bleeding risk in the ANN. Row 1: R, reaction time in TEG; CI, coagulation index in TEG; Plt count, platelet count; PT, prothrombin time; RISK, risk is defined as L (low, ≤13.5 units of blood component), I (intermediate, 13.5−22.5 units of blood component), and H (high, ≥22.5 units). *Arrow*, grow new network icon; *arrowhead*, start learning icon; *circle*, query icon.

The predicted bleeding risk will be shown in the right column: L (low; $N \leq 13.5$), I (intermediate; $N = 14-22$), or H (high; $N \geq 22.5$), where N is the total number of units transfused in 48 h.

8.4 CONCLUSIONS

This chapter discussed two computer programs, designed and implemented by our group, used as decision support tools for the management of bleeding patients:

- Excel file for transfusion dosage based on laboratory results.
- An artificial neural network for predicting blood use in CPB patients.

The development and implementation of our Excel file for transfusion dosage has greatly increased the productivity and efficiency of our coagulation-based hemotherapy service. This is also an excellent teaching tool for our residents to interpret laboratory tests for transfusion purposes.

To the best of our knowledge, our ANN for predicting blood use is the first one developed for postoperative bleeding risk stratification of CPB patients. With promising results, we have started using this neural network to risk stratify our CPB patients, and it has assisted us in predicting postoperative bleeding risk. More data for the training set are being accumulated to further increase the accuracy of this artificial network.

REFERENCES

[1] Huang SP, Nedelcu E, Bai Y, Wahed A, et al. Mobile computing platform with decision-support modules for hemotherapy. Am J Clin Pathol 2014;141:834−40.
[2] Huang SP, Nedelcu E, Bai Y, Wahed A, et al. Postoperative bleeding risk stratification in cardiac pulmonary bypass patients using artificial neural network. Ann Clin Lab Sci 2015;45(2):181−6.

INDEX

Note: Page numbers followed by "*f*" and "*t*" refer to figures and tables, respectively.

Printed in the United States
By Bookmasters